About this Reprint

Prepared for publishing by HP

The Riverside Biographical Series

NUMBER 7

ULYSSES S. GRANT

BY

WALTER ALLEN

U. S. Grant

ULYSSES S. GRANT

BY

WALTER ALLEN

BOSTON AND NEW YORK
HOUGHTON, MIFFLIN AND COMPANY
The Riverside Press, Cambridge
1901

CONTENTS

ULYSSES SIMPSON GRANT

CHAPTER I

OUR NATIONAL MILITARY HERO

Since the end of the civil war in the United States, whoever has occasion to name the three most distinguished representatives of our national greatness is apt to name Washington, Lincoln, and Grant. General Grant is now our national military hero. Of Washington it has often been said that he was "first in war, first in peace, and first in the hearts of his countrymen." When this eulogy was wholly just the nation had been engaged in no war on a grander scale than the war for independence. That war, in the numbers engaged, in the multitude and renown of its battles, in the territory over which its campaigns were extended, in its destruction of life and waste of property, in the magnitude of the interests at stake (but not in the vital importance of the issue), was far inferior to the civil war. It happens quite naturally, as in so many other affairs in this world, that the comparative physical magnitude of the conflicts has much influence in moulding the popular estimate of the rank of the victorious commanders.

Those who think that in our civil war there were other officers in both armies who were Grant's superiors in some points of generalship will hardly dispute that, taking all in all, he was supreme among the generals on the side of the Union. He whom Sherman, Sheridan, Thomas, and Meade saw promoted to be their commander, not only without envy, but with high gratification, under whom they all served with cordial confidence and enthusiasm, cannot have been esteemed by them unfit for the distinction. If these great soldiers then and always acclaimed him worthy to be their leader, it is unbecoming for others, and especially for men who are not soldiers, to contradict their judgment.

Whether he was a greater soldier than General Robert E. Lee, the commander-in-chief of the army of the Confederate States, is a question on which there may always be two opinions. As time passes, and the passions of the war expire, it may be that wise students of military history, weighing the achievements of each under the conditions imposed, will decide that in some respects Lee was Grant's superior in mastery of the art of war. Whether or not this comes about, Lee can never supplant Grant as our national military hero. He fought to destroy the Union, not to save it, and in the end he was beaten by General Grant. However much men may praise the personal virtues and the desperate achievements of the great warrior of the revolt against the Union, they cannot conceal that he was the defeated leader of a lost cause, a cause which, in the chastened judgment of coming time, will appear to all men, as even now it does to most dispassionate patriots, well and fortunately lost.

In the story of Grant's life some things

must be told that are not at all heroic. Much as it might be wished that he had been what Carlyle says a hero should be, a hero at all points, he was not a worshipful hero. Like ourselves all, he was a combination of qualities good and not good. The lesson and encouragement of his life are that in spite of weaknesses which at one time seemed to have doomed him to failure and oblivion, he so mastered himself upon opportune occasion that he was able to prove his power to command great and intelligent armies fighting in a right cause, to obtain the confidence of Lincoln and of his loyal countrymen, and to secure a fame as noble and enduring as any that has been won with the sword.

CHAPTER II

HIS ANCESTRY

THIS hero of ours was of an excellent ancestry. Until lately, most Americans have been careless of preserving their family records. That they were Americans and of a respectable line, if not a distinguished one, for two or three generations back, was as much of family history as interested them, and all they really knew. This was especially true of families which had emigrated from place to place as pioneers in the settlement of the country. Family records were left behind, and in the hard desperate work of life in a new country, where everything depended on individual qualities, and forefathers counted for little in the esteem of men as poor, as independent, and as aspiring as themselves, memories faded and traditions were forgotten. It was esteemed a condition of the equality which was the national boast that no one should take credit to himself on account of distant ancestry. Not until Abraham Lincoln had honored his name by his own nobility did anybody think it worth while to inquire whether his blood was of the strain of the New England Lincolns.

All that was known of the Grants in Ohio was that Jesse, the father of Ulysses, came from Pennsylvania. Jesse himself knew that his father, who died when he was a boy, was Noah Grant, Jr., who came into Pennsylvania from Connecticut, and he had made some further exploration of his genealogical line. But this was more than his neighbors knew or cared to know about the family, until a son demonstrated possession of extraordinary qualities, which set the believers in heredity upon making investigation. The Grants are traced back through Pennsylvania to Connecticut, and from Connecticut to the Massachusetts Bay Colony, where Matthew Grant lived in 1630. He is believed to have come from Scotland, where the Grant clan has been distinguished for centuries on account of its sturdy indomitable traits and its prowess in war. The chiefs of the clan had armorial crests of which the conspicuous emblem was commonly a burning mountain, and the motto some expression of unyielding firmness. In one case it was, "Stand Fast, Craig Ellarchie!" in another, simply "Stand Fast;" in another, "Stand Sure." Sometimes Latin equivalents were used, as "Stabit" and "Immobile." It is said that, as late as the Sepoy rebellion in India, there was a squadron of British troops, composed almost entirely of Scotch Grants, who carried a banner with the motto: "Stand Fast, Craig Ellarchie!"

If it be true that our General Grant came from such stock, his most notable characteristics are no mystery. It was in his blood to be what he was. Ancestral traits reappeared in him with a vigor never excelled. But they had not been quite dormant during the intermediate period. His great-grandfather, Captain Noah Grant, of Windsor (now Tolland), Conn., commanded a company of colonial militia in the French and Indian war, and was killed in the battle of

White Plains in 1776. His grandfather Noah was a lieutenant in a company of the Connecticut militia which marched to the succor of Massachusetts in the beginning of the Revolution. He served, off and on, through the war.

Regarding the circumstances of the removal to Pennsylvania little is known. The home was in Westmoreland County, where Jesse R. Grant was born. Soon afterwards the family went to Ohio. When Jesse was sixteen he was sent to Maysville, Ky., and apprenticed to the tanner's trade, which he learned thoroughly, and made the chief occupation of his life. Soon after he reached his majority he started in business for himself in Ravenna, Portage County, Ohio. In a short time he removed to Point Pleasant, on the Ohio side of the Ohio River, about twenty miles above Cincinnati. Here he lived and prospered for many years, marrying, in 1821, Hannah Simpson, daughter of a farmer of the place in good circumstances. The Simpsons were also of Scotch ancestry, and of stout, self-reliant, industrious, respectable character, like the Grants. Thus in the parents of General Grant were united strains of one of the strong races of the world, — sound in body, mind, and soul, and having in a remarkable degree vital energy, the spirit of independence, and the staying power which enables its possessors to work without tiring, to endure hardships with fortitude, and to accumulate a competence by patient thrift. This last ability General Grant lacked.

These parents, like those of the majority of Americans of the old stock, thought it no dishonor to toil for livelihood, cultivating their souls' health by performance of daily duty in fidelity to God, their country, and their home. Jesse R. Grant had slight opportunities of schooling, but he had no contempt for knowledge. Throughout his life he was a diligent reader of books and newspapers, and was rated a man of uncommon intelligence and of sound judgment in business. He was an entertaining talker, and a newspaper writer and public speaker of local celebrity. Through his early manhood, while he lived in Ohio, he was a farmer, a trader, a contractor for buildings and roads, as well as a tanner. When he reached the age of sixty, having secured a comfortable competence, he retired from active business. In his declining years he removed to Covington, Ky., near Cincinnati. Mrs. Grant was a true helpmate, a woman of refinement of nature, of controlling religious faith, being from her youth an active member of the Methodist Church, of strong wifely and maternal instincts. Her life was centred in her home and family. Both these parents lived to rejoice in the high achievement and station of their son Ulysses.

CHAPTER III

THE PERIOD OF YOUTH

OF such ancestry General Grant was born April 27, 1822, in Point Pleasant, Ohio, and was named Hiram Ulysses Grant. A picture of the house in which he was born shows it to have been a small frame dwelling of primitive character. Its roof, sloping to the road in front, inclosed the two or three rooms that may have been above the ground floor. The principal door was in the middle of the front, and there was one small window on each side of it. Apparently there was a low extension in the rear. This manner of house immediately succeeded the primal log cabins of the Western States, and such houses have sufficed for the happy shelter of large families of strong boys and blooming girls, as sound in body and soul, if not so refined and variously accomplished, as are reared in mansions of more pretension. Love, virtue, industry, and mutual helpfulness made true homes and bred useful citizens.

In the next year his parents removed to the village of Georgetown, Ohio, in Brown County, where the father continued his business of tanner. There young Grant lived until he became a cadet in the Military Academy at West Point. His life was that of other boys of like condition, with few uncommon incidents. Being the eldest of an increasing family, it naturally happened that he was required to perform a share of work for its support, and to bear responsibilities. In his early youth his employment was in the farm work, and this he always preferred. He had a native liking for the open air, and enjoyed the smell of furrows and pastures and woods more than that of reeking hides in their vats. He was fond of all animals, and especially delighted in horses, early demonstrating a surprising power in managing them. He was locally noted for his success in breaking colts, and as a trainer of horses to be pacers, those having this gait being esteemed more desirable for riding, at a time when a large part of all traveling was done on horseback. As General Grant became famous at a comparatively early age, a large crop of stories of his early feats in the subjection and use of horses was cultivated by persons who knew him as a boy. Many of these, doubtless, are entirely credible; few of them are so extraordinary that they might not be true of any clever boy who loved horses and studied their disposition and powers.

He was a lad of self-reliance, fertile in resources, and of good judgment within certain limitations. Before he was fairly in his teens his father intrusted to him domestic and business affairs which required him to go to the city of Cincinnati alone, a two-days' trip. His own account of this period of his life is: "When I was seven or eight years of age I began hauling [driving the team] all the wood used in the house and shops. . . . When about eleven years old, I was strong enough to hold a plow. From that age until seventeen, I did all the work done with horses. . . . While still young, I had visited Cincinnati, forty-five miles away, several times alone; also Maysville, Ky., often, and once Louisville. . . . I did not like to work; but I did as much of it while young as grown men can be hired to do in these days, and attended school at the same time. . . . The rod was freely used there, and I was not exempt from its influence."

But his knowledge of horses, of timber, and of land was better than his knowledge of men. He had no precocious "smartness," as the Yankees name the quality which enables one person to outwit another. His credulity was simple and unsuspecting, at least in some directions. This is illustrated by a story which he has told himself, one which he was never allowed to forget: —

"There was a Mr. Ralston, who owned a colt which I very much wanted. My father had offered twenty dollars for it, but Ralston wanted twenty-five. I was so anxious to have the colt that . . . my father yielded, but said twenty dollars was all the horse was worth, and told me to offer that price. If it was not accepted, I was to offer twenty-two and a half, and if that would not get him, to give the twenty-five. I at once mounted a horse and went for the colt. When I got to Mr. Ralston's house, I said to him: 'Papa says I may offer you twenty dollars for the colt, but, if you won't take that, I am to offer twenty-two and a half, and if you won't take that, to give you twenty-five.'" This naïve bargaining was done when he was eight years old. Some persons have thought it betokens a defect in business acumen which was never fully cured.

He learned his school tasks without great effort. His parents were alive to the advantages of education, and required him to attend all the subscription schools kept in the town. There were no free schools there during his youth. He was twice sent away from home to attend higher schools. It is not recorded that he especially liked study or disliked it. Probably he took it as a part of life, something that had to be done, and did it. He was most apt in mathematics. When he arrived at West Point he was able to pass the not very severe entrance examination without trouble. He seems to have had good native powers of perception, reasoning, and memory. What he learned he kept, but he was never an ardent scholar. He had no enthusiasm for knowledge, nor, indeed, so far as appears, for anything else except horses. He used to fish occasionally, but never hunted. The sportsman's tastes were not his, nor were his social tastes demonstrative. Possibly they may have been restrained in some measure by his mother's strictness of religious principles. He was neither morose nor brooding, — not a dreamer of destiny. He yearned for no star. No instinct of his future achievements made him peculiar among his companions or caused him to hold himself aloof. He exhibited nothing of the young Napoleon's distemper of gnawing pride. He was just an ordinary American boy, with rather less boyishness and rather more sobriety than most, disposed to listen to the talk of his elders instead of that of persons of his own age, and fond of visiting strange places and riding and driving about the country. His work had made him acquainted with the subjects in which grown men were interested. The family life was serious but not severe. Obedience and other domestic virtues were inculcated with fidelity; but he said that he was never scolded or punished at home.

CHAPTER IV

HIS LIFEWORK APPOINTED

WHEN the boy was about seventeen years old he had made up his mind upon one matter, — he would not be a tanner for life. He told his father, possibly in response to some suggestion that it was time for him to quit his aimless occupations and begin his lifework, that he would work in the shop, if he must, until he was twenty-one, but not a day longer. His desire then was to be a farmer, or a trader, or to get an education; but he seems to have had no definite inclination except to escape from the disagreeable tannery. His father treated the matter judiciously, not being disposed to force the boy to learn a business that he would not follow. He was unable to set him up in farming. He had not much respect for the river traders, and may have had little confidence in the boy's ability to thrive in competitions of enterprise and greed.

Without consulting his son, he wrote to one of the United States Senators from Ohio, Hon. Thomas Morris, telling him that there was a vacancy in the district's representation in West Point, and asking that Ulysses might be appointed. He would not write to the congressman from the district, because, although neighbors and old friends, they belonged to different parties and had had a falling out. But the Senator turned the letter over to the Congressman, who procured the appointment, thus healing a breach of which both were ashamed. General Grant gives an account of what happened when this door to an education and a life service was opened before him. His father said to him one day: "'Ulysses, I believe you are going to receive the appointment.' 'What appointment?' I inquired. 'To West Point. I have applied for it.' 'But I won't go,' I said. He said he thought I would, *and I thought so too, if he did.*" The italics are the general's. They make it plain that he did not think it prudent to make further objection when his father had reached a decision.

Little did Congressman Thomas L. Hamer imagine that in doing this favor for his friend, Jesse Grant, he was doing the one thing that would secure remembrance of his name by coming generations. It did not contribute to his immediate popularity among his constituents, for the general opinion was that many brighter and more deserving boys lived in the district, and one of them should have been preferred. Neighbors did not hesitate to shake their heads and express the opinion that the appointment was unwise. Not one of them had discerned any particular promise in the boy. Nor were they unreasonable. He was without other distinctions than of being a strong toiler, good-natured, and having a knack with horses. He had no aspiration for the career of a soldier, in fact never intended to stick to it. Even after entering West Point his hope, he has said, was to be able, by reason of his education, to get "a permanent position in some respectable college," — to become Professor Grant, not General Grant.

In the course of making his appointment,

his name by an accident was permanently changed. When Congressman Hamer was asked for the full name of his protégé to be inserted in the warrant, he knew that his name was Ulysses, and was sure there was more of it. He knew that the maiden name of his friend's wife was Simpson. At a venture, he gave the boy's name as Ulysses Simpson Grant. Grant found it so recorded when he reached the school, and as he had no special fondness for the name Hiram, which was bestowed to gratify an aged relative, he thought it not worth while to go through a long red-tape process to correct the error. There was another Cadet Grant, and their comrades distinguished this one by sundry nicknames, of which "Uncle Sam" was one and "Useless" another.

When he arrived at West Point, in July, 1839, he was not a prepossessing figure of a young gentleman. The rusticity of his previous occupation and breeding was upon him. Seventeen years old, hardly more than five feet tall, but solid and muscular, with no particular charm of face or manner, no special dignity of carriage, he was only a common sort of pleb, modest, good-natured, respectful, companionable but sober-minded, observant but undemonstrative, willing but not ardent, trusty but without high ambitions, — the kind of boy who might achieve commendable success in the academy, or might prove unequal to its requirements, without giving cause of surprise to his associates.

He had no difficulty in passing the examination at the end of his six months' probationary period, which enabled him to be enrolled in the army, and he was never really in danger of dismissal for deficient scholarship. He seems to have made no effort for superior excellence in scholarship, and in some studies his rank was low. Mathematics gave him no trouble, and he says that he rarely read over any of his lessons more than once, which is evidence that he had unusual power of concentrating his attention, the secret of quick work in study. This power and a faithful memory will enable any one to achieve high distinction if he is willing to toil for it. Grant was not willing to toil for it. He gave time to other things, not in the routine prescribed. He pursued a generous course of reading in modern English fiction, including all the works then published of Scott, Bulwer, Marryat, Lever, Cooper, and Washington Irving, and much besides.

The thing for which he was especially distinguished was, as may be surmised, horsemanship. He was esteemed one of the best horsemen of his time at the academy. But this, too, was easy for him. He appears to have been on good terms with his fellows and well liked, but he was not a leader among them. He has said that while at home he did not like to work. It must be judged that his mind was affected by a certain indolence, that he was capable enough when he addressed himself to any particular task, but not self-disposed to exertion. He felt no constant, pricking incitement to do his best; but was content to do fairly well, as well as was necessary for the immediate occasion. One of his comrades in the academy said in later years that he remembered him as "a very uncle-like sort of a youth. . . . He exhibited but little enthusiasm in anything."

He was graduated in 1843, at the age of 21 years, ranking 21 in a class of 39, a little below the middle station. He had grown 6 inches taller while at the academy, standing 5 feet 7 inches, but weighed no more than when he entered, 117 pounds. His physical condition had been somewhat reduced at the end of his term by the wearing effect of a threatening cough. It cannot be said that any one then expected him to do great things. The characteristics of his early youth that have been set forth were persistent. He was older, wiser, more accomplished, better balanced, but in fundamental traits he was still the Ulysses Grant of the farm — hardly changed at all. No more at school than at home was his life vitiated by vices. He was neither profane nor filthy. His temperament was cool and wholesome. He tried to learn to smoke, but was then unable. It is remembered that during the vacation in the middle of his course, spent at home, he steadily declined all invitations to partake of intoxicants, the reason assigned being that he with others had pledged themselves not to drink at all, for the sake of example and help to one of their number whose good resolutions needed such propping. At his graduation he was a man and a soldier. Life, with all its attractions and opportunities, was before. Phlegmatic as he may have been, it cannot be supposed that the future was without beckoning voices and the rosy glamour of hope.

CHAPTER V

LOVE AND WAR

He had applied for an appointment in the dragoons, the designation of the one regiment of cavalry then a part of our army. His alternative selection was the Fourth Infantry. To this he was attached as a brevet second lieutenant, and after the expiration of the usual leave spent at home, he joined his regiment at Jefferson Barracks, St. Louis. Duties were not severe, and the officers entertained much company at the barracks and gave much time to society in the neighborhood. Grant had his saddle-horse, a gift of his father, and took his full share in the social life. A few miles away was the home of his classmate and chum during his last year at the academy, F. T. Dent. One of Dent's sisters was a young lady of seventeen, educated at a St. Louis boarding school. After she returned to her home in the late winter young Grant found the Dent homestead more attractive than ever.

This was the time of the agitation regarding the annexation of Texas, a policy to which young Grant was strongly hostile. About May 1 of the next year, 1844, some of the troops at the barracks were ordered to New Orleans. Grant, thinking his own regiment might go soon, got a twenty-days leave to visit his home. He had hardly arrived when by a letter from a fellow officer he learned that the Fourth had started to follow the Third, and that his belongings had been forwarded. It was then that he became conscious of the real nature of his feeling for Julia Dent. His leave required him to report to Jefferson Barracks, and although he knew his regiment had gone, he construed the orders literally and returned there, staying only long enough to declare his love and learn that it was reciprocated. The secret was not made known to the parents of the young lady until the next year, when he returned on a furlough to see her. For three years longer they were separated, while he was winning honor and promotion. After peace was declared, and the regiment had returned to the States, they were married. She shared all his vicissitudes of fortune until his death. Their life together was one in which wifely faith and duty failed not, nor did he fail to honor and esteem her above all women. Whatever his weaknesses, infidelity in domestic affection was not one of them. In all relations of a personal character he reciprocated trust with the whole tenacity of his nature.

In Louisiana the regiment encamped on high ground near the Sabine River, not far from the old town of Natchitoches. The camp was named Camp Salubrity. In Grant's case, certainly, the name was justified. There he got rid of the cough that had fastened upon him at West Point and had caused fears that he would early fall a victim to consumption. In Louisiana he was restored to perfect, lusty health, fit for any exertion or privation. He was regarded as a modest and amiable lieutenant of no great promise. The regiment was moved to Corpus Christi, a trading and smuggling port. There the army of occupation (of Texas) was slowly collected, consisting of about three thousand men, commanded by General Zachary Taylor. Mexico still claimed this part of Texas, and it was expected that our forces would be attacked. But they were not, and, as the real purpose was to provoke attack, the army was moved to a point opposite Matamoras on the Rio Grande, where a new camp was established and fortified. Previous to leaving Corpus Christi, Grant had been promoted, September 30, 1845, from brevet second lieutenant to full second lieutenant. The advance was made in March, 1846. On the 8th of May the battle of Palo Alto was fought, on the hither side of the Rio Grande, in which Grant had an active part, acquitting himself with credit. On the next day was the battle of Resaca de la Palma, in which he was acting adjutant in place of the officer killed. One consequence of these victories was the evacuation of Matamoras. War with Mexico having been declared, General Taylor's army became an army of invasion.

Volunteers for the war now began coming from the States. In August the movement on Monterey began, and on the 19th of September, Taylor's army was encamped before the city. The battle of Monterey was begun on the 21st, and the desperately defended city was surrendered and evacuated on the 24th. Grant, although then doing quartermaster's duty, having his station with the baggage train, went to the front on the first day, and was a participant in the assault, incurring all its perils, and volunteering for the extremely hazardous duty of a messenger between different parts of the force.

When General Scott arrived at the mouth of the Rio Grande, Grant's regiment was detached from Taylor's army and joined Scott's. He was present and participated in the siege of Vera Cruz, the battle of Cerro Gordo, the assault on Churubusco, the storming of Chapultepec, for which he volunteered with a part of his company, and the battle of Molino del Rey. Colonel Garland, commander of the brigade, in his report of the storming of Chapultepec, said: "Lieutenant Grant, 4th Infantry, acquitted himself most nobly upon several occasions under my own observation." After the battle of Molino del Rey he was appointed on the field a first lieutenant for his gallantry. For his conduct at Chapultepec he was later brevetted a captain, to date from that battle, September 13, 1847. He entered the city of Mexico a first lieutenant, after having been, as he says, in all the engagements of the war possible for any one man, in a regiment that lost more officers during the war than it ever had present in a single engagement.

Perhaps his most notable exploit was during the assault on the gate of San Cosme, under command of General Worth. While reconnoitring for position, Grant observed a church not far away, having a belfry. With another officer and a howitzer, and men to work it, he reached the church, and, by dismounting the gun, carried it to the belfry, where it was mounted again but a few hundred yards from San Cosme, and did excellent service. General Worth sent

Lieutenant Pemberton (the same who in the civil war defended Vicksburg) to bring Grant to him. The general complimented Lieutenant Grant on the execution his gun was doing, and ordered a captain of voltigeurs to report to him with another gun. "I could not tell the general," says Grant, "that there was not room enough in the steeple for another gun, because he probably would have looked upon such a statement as a contradiction from a second lieutenant. I took the captain with me, but did not use his gun."

The American army entered the city of Mexico, September 14, 1847, and this was his station until June, 1848, when the American army was withdrawn from Mexico, peace being established. There was no more fighting. Grant was occupied with his duties as quartermaster, and in making excursions about the country, in which and its people he conceived a warm interest that never changed. Upon returning to his own country he left his regiment on a furlough of four months. His first business was to go to St. Louis and execute his promise to marry Miss Dent. The remainder of this honeymoon vacation was spent with his family and friends in Ohio.

CHAPTER VI

YEARS OF DORMANT POWER

ALTHOUGH he had done excellent service, demonstrating his courage, his good judgment, his resourcefulness, his ability in command, and in the staff duties of quartermaster and commissary, his experience did not kindle in him any new love for his profession, nor any ardor of military glory. He had not revealed the possession of extraordinary talent, nor any spark of genius. He accounted the period of great value to him in his later life, but his heart was never enlisted in the cause for which the war was made. His letters home declared this. When he came to write his memoirs, speaking of the annexation of Texas, he said: "For myself I was bitterly opposed to the measure, and to this day regard the war which resulted as one of the most unjust ever waged by a stronger against a weaker nation. It was an instance of a republic following the bad example of European monarchies in not considering justice in their desire to acquire additional territory. . . . The Southern rebellion was largely the outgrowth of the Mexican war. . . . We got our punishment in the most sanguinary and expensive war of modern times."

But the Mexican war changed Grant's plan of life. While he was at Jefferson Barracks he had applied for a place as instructor of mathematics at West Point, and had received such encouragement that he devoted much time to reviewing his studies and extending them, giving more attention to history than ever before. After the war the notion of becoming a college professor appears to have left him. He regarded himself as bound to the service for the rest of his days. It was not so much his choice as his lot, and he accepted it, not because he relished it, but because he discovered no way out of it. This illustrates a negative trait of his character remarked throughout his career. He was never a pushing man.

He had no self-seeking energy. The work that was assigned to him he did as well as he could; but he had little art to recommend himself in immodest ways. He had not the vanity to presume that he would certainly succeed in strange enterprises. He shrank from the personal hostilities of ambition.

Then followed a long period of uneventful routine service in garrisons at Detroit and at Sackett's Harbor, until in the summer of 1852 his regiment was sent to the Pacific coast via the Panama route. The crossing of the isthmus was a terrible experience, owing to the lack of proper provision for it and to an epidemic of cholera. The delay was of seven weeks' duration, and about one seventh of all who sailed on the steamer from New York died on the isthmus of disease or of hardships. Lieutenant Grant, however, had no illness, and exhibited a humane devotion to the necessities of the unfortunate, civilians as well as soldiers. His company was destined to Fort Vancouver, in Oregon Territory, where he remained nearly a year, until, in order, he received promotion to a captaincy in a company stationed at Humboldt Bay in California. Here he remained until 1854, when he resigned from the army, because, as he says, he saw no prospect of being able to support his family on his pay, if he brought them — there were then two children — from St. Louis, where Mrs. Grant had remained with her family since he left New York. His resignation took effect, following a leave of absence, July 31, 1854.

There was another cause, as told in army circles, for his resignation. He had become so addicted to drink that his resignation was required by his commanders, who held it for a time to afford him an opportunity to retrieve his good fame if he would; but he was unable. Through what temptation he fell into such disgrace is not clearly known. But garrison posts are given to indulgences which have proved too much for many an officer, no worse than his fellows, but constitutionally unable to keep pace with men of different temperament. It might be thought that Grant was one unlikely to be easily affected; but the testimony of his associates is that he was always a poor drinker, a small quantity of liquor overcoming him.

He was now thirty-two years old, a husband and father, discharged from the service for which he had been educated, and without means of livelihood. His wife fortunately owned a small farm near St. Louis, but it was without a dwelling house. He had no means to stock it. He built a humble house there by his own hard labor. He cut wood and drew it to St. Louis for a market. In this way he lived for four years, when he was incapacitated for such work by an attack of fever and ague lasting nearly a year. There is no doubt that the veteran and his family experienced the rigors of want in these years; no question that neither his necessities nor his duties saved him from being sometimes overcome by his baneful habit.

In the fall of 1858 the farm was sold. Grant embarked in the real estate agency business in St. Louis, and made sundry unsuccessful efforts to get a salaried place under the city government. But his fortunes did not improve. Finally in desperation he went in 1860 to his father for assistance. His father had established two younger sons in a hide and leather business in Galena, Ill. Upon consultation they agreed to employ Ulysses as a clerk and helper, with the understanding that he should not draw more than $800 a year. But he had debts in St. Louis, and to cancel these almost as much more had to be supplied to him the first year. His father has told that the advance was repaid as soon as he began earning money in the civil war.

In Galena he was known to but few. Ambition for acquaintance seemed to have died in him. He was the victim of a great humiliation and was silent. He avoided publicity. He was destitute of presumption. What brighter hopes he cherished were due to his father's purpose to make him a partner with his brothers. He heard Lincoln and Douglas when they canvassed the State, and approved of the argument of the former rather than of the other. He had voted for

Buchanan in 1856, his only vote for a President before the war. In 1860 he had not acquired a right to vote in Illinois.

These thirteen quiet years of Grant's life are not of account in his public career, but they are a phase of experience that left its deep traces in the character of the man. He was changed, and ever afterwards there was a tinge of melancholy and a haunting shadow of dark days in his life that could not be escaped. Nor in the pride and power of his after success did he completely conquer the besetting weakness of his flesh. The years from twenty-six to thirty-nine in the lives of most men who ever amount to anything are years of steady development and acquisition, of high endeavor, of zealous, well-ordered upward progress, of growth in self-mastery and outward influence, of firm consolidation of character. These conditions are not obvious in the case of General Grant. Had he died before the summer of 1861, being nearly forty years of age, he would have filled an obscure grave, and those to whom he was dearest could not have esteemed his life successful, even in its humble scope. He had not yet found his opportunity: he had not yet found himself.

CHAPTER VII

THE SUMMONS OF PATRIOTISM

THE tide of patriotism that surged through the North after the fall of Fort Sumter in April, 1861, lifted many strong but discouraged men out of their plight of hard conditions and floated them on to better fortune. Grant was one of these. At last he found reason to be glad that he had the education and experience of a soldier.

On Monday, April 15, 1861, Galena learned that Sumter had fallen. The next day there was a town meeting, where indignation and devotion found utterance. Over that meeting Captain Grant was called to preside, although few knew him. Elihu B. Washburn, the representative of the district in Congress, and John A. Rawlins, a rude, self-educated lawyer, who had been a farmer and a charcoal burner, made passionate, fiery speeches on the duty of every man to stand by the flag. At the close of that meeting Grant told his brothers that he felt that he must join the army, and he did no more work in the shop. How clearly he perceived the meaning of the conflict was shown in a letter to his father-in-law, wherein he wrote: "In all this I can see but the doom of slavery."

He was offered the captaincy of the company formed in Galena, and declined it, although he aided in organizing and drilling the men, and accompanied them to the state capital, Springfield. As he was about starting for home, he was asked by Governor Richard Yates to assist in the adjutant-general's office, and soon he was given charge of mustering in ten regiments that had been recruited in excess of the quota of the State, under the President's first call, in preparation for possible additional calls. His knowledge of army forms and methods was of great service to the inexperienced state officers.

Later, but without wholly severing his connection with the office, he returned home, and wrote a letter to the adjutant-general of the regular army, at Washington, briefly setting forth his former service, and very respectfully tendering his service "until the close of the war in such capacity as may be offered," adding, that with his experience he felt that he was "competent to command a regiment, if the President should see fit to intrust one to him." The letter brought no reply. He went to Cincinnati and tried, unsuccessfully, to see General McClellan, whom he had known at West Point and in Mexico, hoping that he might be offered a place on his staff. While he was absent Governor Yates appointed him colonel of the Twenty-First Regiment of Illinois Infantry, then in camp near Springfield, his commission dating from June 15. It was a thirty-day regiment, but almost every member reënlisted for three years, under the President's second call. Thus, two months after the breaking out of the war, he was again a soldier with a much higher commission than he had ever held, higher than would have come to him in regular order had he remained in the army.

At Springfield he was in the centre of a great activity and a great enthusiasm. He met for the first time many leading men of the State, and became known to them. Their personality did not overwhelm him, famous and influential as many of them were, nor did he solicit from them any favor for himself. His desire was to be restored to the regular army rather than to take command of volunteers. When the sought-for opportunity did not appear, he accepted the place that was offered, a place in which he was needed; for the first colonel, selected by the regiment itself, had already by his conduct lost their confidence. They exchanged him for Grant with high satisfaction.

CHAPTER VIII

FROM SPRINGFIELD TO FORT DONELSON

THE regiment remained at the camp, near Springfield, until the 3d of July, being then in a good state of discipline, and officers and men having become acquainted with company drill. It was then ordered to Quincy, on the Mississippi River, and Colonel Grant, for reasons of instruction, decided to march his regiment instead of going by the railroad. So began his advance, which ended less than four years later at Appomattox, when he was the captain of all the victorious Union armies, — holding a military rank none had held since Washington, — and a sure fame with the great captains of the world's history. The details of this wonderful progress can only be sketched in this little volume. It was not without its periods of gloom, and doubt, and check; but, on the whole, it was steadily on and up.

His orders were changed at different times, until finally he was directed to proceed with all dispatch to the relief of an Illinois regiment, reported to be surrounded by rebels near Palmyra, Mo. Before the place was reached, the imperiled regiment had delivered itself by retreating. He next expected to give battle at a place near the little town of Florida, in Missouri. As the regiment toiled over the hill beyond which the enemy was supposed to be waiting for him, he "would have given anything to be back in Illinois." Never having had the responsibility of command in a fight, he really distrusted his untried ability. When the top of the hill was reached, only a deserted camp appeared in front. "It occurred to me at once that Harris had been as much afraid of me as I had been of him. . . . From that event to the close of the war," he says in his book, "I never experienced trepidation upon confronting an enemy, though I always felt more or less anxiety. I never forgot that he had as much reason to fear my forces as I had [to fear] his."

On August 7 he was appointed by the President a brigadier-general of volunteers, upon the unanimous recommendation of the congressmen from Illinois, most of whom were unknown to him. He had not won promotion by any fighting; but generals were at that time made with haste to meet exigent requirement, a proportional number being selected from each loyal State. Among those whom General Grant appointed on his staff was John A. Rawlins, the Galena lawyer, who was made adjutant-general, with the rank of captain, and who as long as he lived continued near Grant in some capacity, dying while serving as Secretary of War in the first term of Grant's presidency. He was an officer of high ability and personal loyalty. He alone had the audacity to interpose a resolute no, when his chief was disposed to over-indulgence in liquor. He did not always prevent him, but it is doubtful whether Grant would not have fallen by the way without the constant, imperative watchfulness of his faithful friend. There were times when both army and people were impatient with him, not wholly without reason. Nothing saved him then but President Lincoln's confidence and charity. The reply to all complaints was: "This man fights; he cannot be spared."

In the last days of August, having been occupied, meantime, in reducing to order distracted and disaffected communities in Missouri, he was assigned to command of a military district embracing all southwestern Missouri and southern Illinois. He established his headquarters at Cairo, early in September, and from there he promptly led an expedition that forestalled the hostile intention of seizing Paducah, a strategical point at the mouth of the Tennessee River. This was his first important military movement, and it was begun upon his own initiative. His first battle was fought at Belmont, Mo., opposite Columbus, Ky., on the Mississippi River, on November 7, 1861. Grant, in command of a force of about 3000 men, was demonstrating against Columbus, held by the enemy. Learning that a force had been sent across the river to Belmont, he disembarked his troops from their transports and attacked. The men were under fire for the first time, but they drove the enemy and captured the camp. They came near being cut off, however, through the inexperience and silly recklessness of subordinate officers. By dint of hard work and great personal risk on the part of their commander, they were got safely away. It was an all-day struggle, during which General Grant had a horse shot under him, and made several narrow escapes, being the last man to reëmbark. The Union losses were 485 killed, wounded, and missing. The loss of the enemy was officially reported as 632. This battle was criticised at the time as unnecessary; but General Grant always asserted the contrary. The enemy was prevented from detaching troops from Columbus, and the national forces acquired a confidence in themselves that was of great value ever afterwards. Grant's governing maxim was, to strike the enemy whenever possible, and keep doing it.

From the battle of Belmont until February, 1862, there was no fighting by Grant's army. Troops were concentrated at Cairo for future operations — not yet decided upon. Major-General H. W. Halleck superseded General Fremont in command of the department of Missouri. Halleck was an able man, having a high reputation as theoretical master of the art of war, one of those who put a large part of all their energy into the business of preparing to do some great task, only to find frequently, when they are completely ready, that the occasion has gone by. When he was first approached with a proposition to capture Forts Henry and Donelson, the first on the Tennessee River, the other on the Cumberland River, where the rivers are only a few miles apart near the southern border of Kentucky, he thought that it would require an army of "not less than 60,000 effective men," which could not be collected at Cairo "before the middle or last of February."

Early in January General Grant went to St. Louis to explain his ideas of a campaign against these forts to Halleck, who told him his scheme was "preposterous." On the 28th he ventured again to suggest to Halleck by telegraph that, if permitted, he could take and hold Fort Henry on the Tennessee. His application was seconded by flag officer Foote of the navy, who then had command of several gunboats at Cairo. On February 1, he received instructions to go ahead, and the expedition, all preparations having been made beforehand, started the next day, the gunboats and about 9000 men on transports going up the Ohio and the Tennessee to a point a few miles below Fort Henry. After the troops were disembarked the transports went back to Paducah for the remainder of the force of 17,000 constituting the expeditionary army. The attack was made on the 6th, but the garrison had evacuated, going toward Fort Donelson, to escape the fire of the gunboats. General Tilghman, commanding the fort, his staff, and about 120 men were captured, with many guns and a large quantity of stores. The principal loss on the Union side was the scalding of 29 men on the gunboat Essex by the explosion of her boiler, pierced by a shell from the fort.

Grant had no instructions to attack Fort Donelson, but he had none forbidding him to do it. He straightway moved nearly his whole force over the eleven miles of dreadful roads, and on the 12th began investing the stronghold, an earthwork inclosing about 100 acres, with outworks on the land and water sides, and defended by more than 20,000 men commanded by General Floyd, who had been President Buchanan's Secretary of War. The investing force had its right near the river above the fort. The weather was alternately wet and freezing cold. The troops had no shelter, and suffered greatly. On the 14th, without serious opposition, the investment was completed. At three o'clock in the afternoon of the 14th, flag officer Foote began the attack, the fleet of gunboats steaming up the river and firing as rapidly as possible; but several were disabled by the enemy's fire, and all had to fall back before nightfall. The enemy telegraphed to Richmond that a great victory had been achieved.

On the next day, Grant, riding several miles to the river, met Foote on his gunboat, to which he was confined by a wound received the day before. Returning, he found that a large force from the fort had made a sortie upon a part of his line, but had been driven back after a severe contest. It was found that the haversacks of the Confederates left on the field contained three days' rations. Instantly, Grant reasoned that the intention was not so much to drive him away as to break through his line and escape. He ordered a division that had not been engaged to advance at once, and before night it had established a position within the outer lines of defense. Surrender or capture the next day was the fate of the Confederates.

During the night General Floyd and General Pillow, next in command, and General Forest made their escape with about 4000 men. Before light the next morning, General Grant received a note from General S. B. Buckner, who was left in command of the fort, suggesting the appointment of commissioners to agree upon terms of capitulation, and meanwhile an armistice until noon. To this note General Grant sent the curt reply: "No terms except an unconditional and immediate surrender can be accepted. I propose to move immediately upon your works." General Buckner sent back word that he was compelled by circumstances "to accept the ungenerous and unchivalrous terms" which had been proposed.

This victory electrified the whole North, then greatly in need of cheer. General Grant became the hero of the hour. His name was honored and his exploit lauded from one end of the country to the other. It was not yet a year since he had been an obscure citizen of an obscure town. Already many regarded him as the nation's hope. A phrase from his note to General Buckner was fitted to his initials, and he was everywhere hailed as "Unconditional Surrender" Grant.

In this campaign he first revealed the peculiar traits of his military genius, clear discernment of possibilities, comprehension of the requirements of the situation, strategical instinct, accurate estimate of the enemy's motive and plan, sagacious promptness of action in exigencies, staunch resolution, inspiring energy, invincible poise. For his achievement he was promoted to be a major-general of volunteers. He had found himself now.

CHAPTER IX

SHILOH, CORINTH, IUKA

On the 4th of March, sixteen days after his victory, he was in disgrace. General Halleck ordered him to turn over the command of the army to General C. F. Smith and to remain himself at Fort Henry. This action of Halleck was the consequence partly of accidents which had prevented communication between them and caused Halleck to think him insubordinate, partly of false reports to Halleck that Grant was drinking to excess, partly of Halleck's dislike of Grant, — a temperamental incapacity of appreciation. After Donelson he issued a general order of congratulation of Grant and Foote for the victory, but he sent no personal congratulations, and reported to Washington that the victory was due to General Smith, whose promotion, not Grant's, he recommended. As to the reports of Grant's drinking, they were decisively contradicted by Rawlins, to whom the authorities in Washington applied for information. He asserted that Grant had drunk no liquor during the campaign except a little, by the surgeon's prescription, on one occasion when attacked by ague. The fault of failing to report his movements and to answer inquiries was later found to be due to a telegraph operator hostile to the Union cause, who did not forward Grant's reports to Halleck nor Halleck's orders to Grant.

Grant's mortification was intense. Since the fall of Donelson he had been full of activities. The enemy had fallen back, his first line being broken, and Grant was scheming to follow up his advantage by pushing on through Tennessee, driving the discouraged Confederate forces before him. He had visited Nashville to confer with General Buell, who had reached that city, and it was on his return that he received Halleck's dispatch of removal. For several days he was in dreadful distress of mind, and contemplated resigning his commission. It seemed as if Fate had cut off his career just as it had gloriously begun. But he made no public complaint. He obeyed orders and waited at Fort Henry. To some of his friends he said that he would never wear a sword again. But on the 13th he was restored to command. Halleck became aware of the facts, and made a report vindicating Grant's conduct, of which he sent him a copy. It was not until after the war that Grant learned that Halleck's previous reports had caused his degradation.

His first battle after restoration to command was an unfortunate one in the beginning, but was turned into a victory. He was advancing on Corinth, Miss., a railroad centre of the Southwest, where a large Confederate army under General Albert Sidney Johnston was collecting. All the available Union forces in the West were gathering to meet it. Grant had selected Pittsburg Landing on the Tennessee River, twenty miles from Corinth, as the place for landing his forces, and Hamburg Landing, four miles up the river, as the starting point for Buell's army in marching on Corinth. Buell was hastening to the rendezvous, coming through Tennessee with a large force. On the 4th of April Grant's horse fell while he was reconnoitring at night, and the general's leg was badly bruised but not broken.

Expecting to make an offensive campaign and meet the enemy at Corinth, he had not enjoined intrenchment of the temporary camp. So great was the confidence that Johnston would await attack that the enemy's proximity in force was discovered too late. Johnston led his whole army out of Corinth, and early on the morning of the 6th of April surprised Sherman's division encamped at Shiloh, three miles from Pittsburg Landing, attacking with a largely superior force. The battle raged all day, with heavy losses on both sides, the Union army being gradually forced back to Pittsburg Landing. Five divisions were engaged, three of them composed of raw troops, and many regiments were in a demoralized condition at night.

On the next day the Union army, reinforced by Buell's 20,000 men, advanced, attacking the enemy early in the morning, with furious determination. The Confederate forces, although weakened, were determined not to lose the advantage gained, and fought with desperate stubbornness. But it was in vain. A necessity of vindicating their courage was felt by officers and men of the Union Army. They had fully recovered from the effects of the surprise, and pressed forward with zealous assurance. Before the day was done Grant had won the field and compelled a disorderly retreat. In this battle the commander of the Confederate army, General Albert Sidney Johnston, was killed in the first day's fighting, the command devolving on General G. T. Beauregard. On the first day the Union forces on the field numbered about 33,000 against the enemy's above 40,000. On the second day the Union forces were superior. The Union losses in the two days were 1754 killed, 8408 wounded, and 2885 missing; total 13,047. Beauregard reported a total loss of 10,694, of whom 1723 were killed. General Grant says that the Union army buried more of the enemy's dead than is here reported in front of Sherman's and McClernand's divisions alone, and that the total number buried was estimated at 4000.

The battles of Shiloh and Pittsburg Landing together constitute one of the critical conflicts of the long war. Had the Confederate success of the first day been repeated and completed on the second day, it would have been difficult, if not impossible, to prevent the enemy from possessing Tennessee and a large part of Kentucky.

After this battle General Halleck came to Pittsburg Landing and took command of all the armies in that department. Although General Grant was second in command, he was not in General Halleck's confidence, and was contemptuously disregarded in the direction of affairs. Halleck proceeded to make a safe campaign against Corinth by road-building and parallel intrenchments. He got there and captured it, indeed, having been a month on the way, but the rebel army, with all its equipments, guns, and stores, had escaped beforehand. Grant's position was so embarrassing that during Halleck's advance he made several earnest applications to be relieved. Halleck would not let him go, apparently thinking that he needed to be instructed by an opportunity of observing how a great soldier made war. What Grant really learned was how not to make war.

After the fall of Corinth he was permitted to make his headquarters at Memphis, while Halleck proceeded to construct defensive works on an immense scale. But in July Halleck was appointed commander-in-chief of all the armies, with his headquarters in Washington, and Grant returned to Corinth. He was the ranking officer in the department, but was not formally assigned to the command until October. The intermediate time was spent, for the most part, in defensive operations in the enemy's country, the great army that entered Corinth having been scattered east, north, and west to various points. Two important battles were fought, by one of which an attempt to retake Corinth was defeated. The other was at

Iuka, in Mississippi, where a considerable Confederate force was defeated.

In this period the energy and resourcefulness of General Grant were conspicuous, although nothing that occurred added largely to his reputation. He was, however, gathering stores of useful experience while operating in the heart of the enemy's country, where every inhabitant, except the negroes, was hostile. Both of the battles mentioned above were nearly lost by failure of his subordinates to render expected service according to orders ; but he suffered no defeat. The service was wearing, but he was equal to all demands made upon him.

CHAPTER X

VICKSBURG

Vicksburg had long been the hard military problem of the Southwest. The city, which had been made a fortress, was at the summit of a range of high bluffs, two hundred and fifty feet above the east bank of the Mississippi River, near the mouth of the Yazoo. It was provided with batteries along the river front and on the bank of the Yazoo to Haines's Bluff. A continuous line of fortifications surrounded the city on the crest of the hill. This hill, the slopes of which were cut by deep ravines, was difficult of ascent in any part in the face of hostile defenders. The back country was swampy bottom land, covered with a rank growth of timber, intersected with lagoons and almost impassable except by a few rude roads. The opposite side of the river was an extensive wooded morass.

In May, 1862, flag officer Farragut, coming up the Mississippi from New Orleans, had demanded the surrender of the city and been refused. In the latter part of June he returned with flag officer Porter's mortar flotilla and bombarded the city for four weeks without gaining his end. In November, 1862, General Grant started with an army from Grand Junction, intending to approach Vicksburg by the way of the Yazoo River and attack it in the rear. But General Van Dorn captured Holly Springs, his depot of supplies, and the project was abandoned.

The narration, with any approach to completeness, of the story of the campaign against Vicksburg would require a volume. It was a protracted, baffling, desperate undertaking to obtain possession of the fortifications that commanded the Mississippi River at that point. Grant was not unaware of the magnitude of the work, nor was he eager to attempt it under the conditions existing. He believed that, in order to their greatest efficiency, all the armies operating between the Alleghanies and the Mississippi should be subject to one commander, and he made this suggestion to the War Department, at the same time testifying his disinterestedness by declining in advance to take the supreme command himself. His suggestion was not immediately adopted. On the 22d of December, 1862, General Grant, whose headquarters were then at Holly Springs, reorganized his army into four corps, the 13th, 15th, 16th, and 17th, commanded respectively by Major-Generals John A. McClernand, William T. Sherman, S. A. Hurlbut, and J. B. McPherson. Soon afterwards he established his headquarters at Memphis, and in January began the move on Vicksburg, which, after immense labors and various failures of plans, resulted in the surrender of that fortress on July 4, 1863.

He first sent Sherman, in whose enterprise and ability to take care of himself he had full confidence, giving him only general instructions. Sherman landed his army on the east side of the river, above Vicksburg, and made a direct assault, which proved unsuccessful, and he was compelled to reëmbark his defeated troops. The impracticability of successful assault on the north side was then accepted. General McClernand's corps on the 11th of January, aided by the navy under Admiral Porter, captured Arkansas Post on the White River, taking 6000 prisoners, 17 guns, and a large amount of military stores.

On the 17th, Grant went to the front and had a conference with Sherman, McClernand, and Porter, the upshot of which was a direction to rendezvous on the west bank in the vicinity of Vicksburg. McClernand was disaffected, having sought at Washington the command of an expedition against Vicksburg and been led to expect it. He wrote a letter to Grant so insolent that the latter was advised to relieve him of all command and send him to the rear. Instead of doing so, he gave him every possible favor and opportunity; but months afterwards, in front of Vicksburg, McClernand was guilty of a breach of discipline which could not be overlooked, and he was deprived of his command.

Throughout the war Grant was notably considerate and charitable in respect of the mistakes and the temper of subordinates if he thought them to be patriotic and capable. His rapid rise excited the jealousy and personal hostility of many ambitious generals. Of this he was conscious, but he did not suffer himself to be affected by it so long as there was no failure in duty. The reply he made to those who asked him to remove McClernand revealed the principle of his action: "No. I cannot afford to quarrel with a man whom I have to command."

The Union army, having embarked at Memphis, was landed on the west bank of the Mississippi River, and the first work undertaken was the digging of a canal across a peninsula that would allow passage of the transports to the Mississippi below Vicksburg, where they could be used to ferry the army across the river, there being higher ground south of the city from which it could be approached more easily than from any other point. After weeks of labor, the scheme had to be abandoned as impracticable. Then various devices for opening and connecting bayous were tried, none of them proving useful. The army not engaged in digging or in cutting through obstructing timbers was encamped along the narrow levee, the only dry land available in the season of flood. Thus three months were seemingly wasted without result. The aspect of affairs was gloomy and desperate.

The North became impatient and began grumbling against the general, doubting his ability, even clamoring for his removal. He made no reply, nor suffered his friends to defend him. He simply worked on in silence. Stories of his incapacity on account of drinking were rife, and it may have been the case that under the dreary circumstances and intense strain he did sometimes yield to this temptation. But he never yielded his aim, never relaxed his grim purpose. Vicksburg must fall. As soon as one plan failed of success another was put in operation. When every scheme of getting the vessels through the by-ways failed, one thing remained, — to send the gunboats and transports past Vicksburg by the river, defying the frowning batteries and whatever impediments might be met. Six gunboats and several steamers ran by the batteries on the night of April 16th, under a tremendous fire, the river being lighted up by burning houses on the shore. Barges and flatboats followed on other nights. Then Grant's way to reach Vicksburg was found; but it was not an easy one, nor unopposed. A place of landing on the east side was to be sought. The navy failed to silence the Confederate batteries at Grand Gulf, twenty miles below Vicksburg, so that a landing could be effected there, and the fleet ran past it, as it had run by Vicksburg. Ten miles farther down the river a landing place was found at Bruinsburg. By daylight, on the 1st of May, McClernand's corps and part of McPherson's had been ferried across, leaving behind all impedimenta, even the officers' horses, and fighting had already begun in rear of Port Gibson, about eight miles from the landing. The enemy made a desperate stand, but was defeated with heavy loss. Grand Gulf was evacuated that night, and the place became thenceforth a base of operations. Grant had defeated the enemy's calculation by the celerity with which he had transferred a large force. He slept on the ground with his soldiers, without a tent or even an overcoat for covering.

General Joseph E. Johnston had superseded General Beauregard in command of all the Confederate forces of the Southwest. His business was to succor General Pemberton and drive Grant back into the river. Sherman with his corps joined Grant on the 8th. Jackson, the capital of Mississippi, a Confederate railroad centre and depot of supplies, was captured on the 14th, the defense being made by Johnston himself. Then Pemberton's whole army from Vicksburg, 25,000 men, was encountered, defeated, and forced to retire into the fortress, after losing nearly 5000 men and 18 guns. On the 18th of May Grant's army reached Vicksburg and the actual siege began.

Since May 1, Grant had won five hard battles, killed and wounded 5200 of the enemy, captured 40 field guns, nearly 5000 prisoners, and a fortified city, compelled the abandonment of Grand Gulf and Haines's Bluff, with their 20 heavy guns, destroyed all the railroads and bridges available by the enemy, separated their armies, which altogether numbered 60,000 men, while his own numbered but 45,000, and had completely invested Vicksburg. It was an astonishing exhibition of courage, energy, and military genius, calculated to confound his critics and reëstablish him in the confidence of the people. It has been said that there is nothing in history since Hannibal invaded Italy that is comparable with it.

The incidents of the siege, abounding in difficult and heroic action, including an early unsuccessful assault, must be passed over. Preparations had been made and directions given for a general assault on the works on the morning of July 5. But on the 3d General Pemberton sent out a flag of truce asking, as Buckner did at Donelson, for the appointment of commissioners to arrange terms of capitulation. Grant declined to appoint commissioners or to accept any terms but unconditional surrender, with humane treatment of all prisoners of war. He, however, offered to meet Pemberton himself, who had been at West Point and in Mexico with him, and confer regarding details. This meeting was held, and on the 4th of July Grant took possession of the city. The Confederates surrendered about 30,000 men, 172 cannon, and 50,000 small arms, besides military stores; but there was little food left. Grant's losses during the whole campaign were 1415 killed, 7395 wounded, 453 missing. When the paroled prisoners were ready to march out, Grant ordered the Union soldiers "to be orderly and quiet as these prisoners pass," and "to make no offensive remarks."

This great victory was coincident with the repulse of Lee at Gettysburg, and the effect of the two events was a wonderfully inspiriting influence upon the country. President Lincoln wrote to General Grant a characteristic letter "as a grateful acknowledgment of the almost inestimable service you have done the country." In it he said: "I never had any faith, except a general hope that you knew better than I, that the Yazoo Pass expedition and the like could succeed. When you got below and took Port Gibson, Grand Gulf, and vicinity, I thought you should go down the river and join General Banks [besieging Port Hudson]; and when you turned northward, east of the Big Black, I feared it was a mistake. I now wish to make a personal acknowledgment that you were right and I was wrong."

Port Hudson surrendered to General Banks, to whom Grant sent reinforcements as soon as Vicksburg fell, on the 8th of July, with 10,000 more prisoners and 50 guns. This put the Union forces in possession of the Mississippi River all the way to the Gulf.

Grant now appeared to the nation as the foremost hero of the war. The disparagements and personal scandals so rife a few months before were silenced and forgotten. He was believed to be invincible. That he never boasted, never publicly resented criticism, never courted applause, never quarreled with his superiors, but endured, toiled, and fought in calm fidelity, consulting chiefly with himself, never wholly baffled, and always triumphant in the end, had shown the nation a man of a kind the people had longed for and in whom they proudly rejoiced. The hopes to which Donelson had given birth were confirmed in the hero of Vicksburg, who was straightway made a major-general in the regular army, from which, when a first lieutenant, he had resigned nine years before.

CHAPTER XI

NEW RESPONSIBILITIES — CHATTANOOGA

HALLECK, issuing orders from Washington, proceeded to disperse Grant's army hither and yon as he thought fractions of it to be needed. Grant wanted to move on Mobile from Lake Pontchartrain, but was not permitted to do it. Having gone to New Orleans in obedience to a necessity of conference with General Banks, he suffered a severe injury by the fall of a fractious horse, as he was returning from a review of Banks's army. For a long time he was unconscious. As soon as he could be moved he was taken on a bed to a steamer. For several days after reaching Vicksburg he was unable to leave his bed. Meantime he was repeatedly called upon to send reinforcements to Rosecrans, in Chattanooga, to which place the latter had retreated after the repulse of his army at Chickamauga,

September 19 and 20. On October 3, Grant was directed to go to Cairo and report by telegraph to the Secretary of War as soon as he was able to take the field. He started on the same day, ill as he still was. On arriving in Cairo he was ordered to proceed to Louisville. He was met at Indianapolis by Secretary Stanton, whom he had never before seen, and they proceeded together.

On the train Secretary Stanton handed him two orders, telling him to take his choice of them. Both created the military division of the Mississippi, including all the territory between the Alleghanies and the Mississippi River, north of General Banks's department, and assigning command of it to Grant. One order left the commanders of the three departments, the Ohio, the Cumberland, and the Tennessee, as they were, the other relieved General Rosecrans, commanding the Army of the Cumberland, and assigned Gen. George H. Thomas to his place. General Grant accepted the latter. This consolidation was a late compliance with his earnest, unselfish counsel given before the Vicksburg campaign. Its wisdom had become apparent.

The centre of interest and anxiety now was Chattanooga, in East Tennessee, near the border of Georgia. The Confederates had been striving to retrieve the ground lost, since the fall of Fort Henry, by pushing northward in this direction. Halleck's dispersion of forces had sent Buell to this section, and Buell had been superseded by Rosecrans, a zealous and patriotic but unfortunate commander. The repulse at Chickamauga might have proved disastrous to his army but for the splendid behavior of the division under General Thomas, an officer not unlike Grant in the mould of his military talent, who there earned the sobriquet, "The Rock of Chickamauga."

The army of Rosecrans had been gathered again at Chattanooga, where it was confronted by Bragg, whose force surrounded it in an irregular semicircle from the Tennessee River to the river again, occupying Missionary Ridge on one flank and Lookout

Mountain on the other, with its centre where these two ridges come nearly together. Chattanooga was in the valley between, near the centre of which, behind the town, was an elevation, Orchard Knob, held by the enemy. Bragg commanded the river and the railroads. The route for supplies was circuitous, inadequate, and insecure, over mountain roads that had become horrible. Horses and mules had perished by thousands. The soldiers were on half rations. Word came to Grant in Louisville, that Rosecrans was contemplating a retreat. He at once issued an order assuming his new command, notified Rosecrans that he was relieved, and instructed Thomas to hold the place at all hazards until he reached the front.

Still so lame that he could not walk without crutches, and had to be carried in arms over places where it was not safe to go on horseback, he left Louisville on the 21st of October, and reached Chattanooga on the evening of the 23d. Then began a work of masterly activity and preparation, in which his genius again asserted its supreme quality. Sherman with his army was ordered to join Grant. In five days the river road to Bridgeport was opened, the enemy being driven from the banks, two bridges were built, and Hooker's army added to his force. The enemy, having a much superior force, and assuming the surrender of the Army of the Cumberland to be only a question of time and famine, sent Longstreet with 15,000 men to reinforce the army of Johnston, holding Burnside in Knoxville, to the relief of whom the enemy supposed Sherman to be marching. Grant waited for Sherman, who was coming on between Longstreet and Bragg. All general orders for the battle were prepared in advance, except their dates. Sherman reached Chattanooga on the evening of the 15th, and with Grant inspected the field on the 16th. Sherman's army, holding the left, was to cross to the south side of the river and assail Missionary Ridge. Hooker, on the right, was to press through from Lookout valley into Chattanooga valley. Thomas, in the centre, was to press forward through the valley and strike the enemy's centre while his wings were thus fully engaged, or as soon as Hooker's support was available.

The battle began on the afternoon of October 23. Orchard Knob, in the centre of the great amphitheatre, was attacked and captured, and became the Union headquarters. On the 24th Sherman crossed the river and established his army on the north end of Missionary Ridge. On the morning of the same day Hooker assailed Lookout Mountain, and after a long climbing fight, lasting far into the night, secured his position; and the enemy, who had occupied the mountain, retreated across the valley at its upper end to Missionary Ridge. Grant's forces were now in touch from right to left. Everything so far had gone well.

Early on the next morning Sherman opened the attack. The ridge in his front was exceedingly favorable for defense, and during the whole night the enemy had been at work strengthening the position. Sherman's first assault failed, but he continued pressing the enemy with resolution, although making little progress. From Grant's place on Orchard Knob he watched the struggle. At three o'clock he saw Sherman's right repulsed. Then he gave to Thomas, standing at his side, the order to advance. Six guns were fired as a signal, and the Army of the Cumberland moved forward in splendid array to avenge Chickamauga. The immediate purpose was to carry the rebel rifle pits at the foot of the Ridge. This done, the soldiers were subjected to a galling fire from the line 800 feet above them. As by inspiration, they rushed on, climbing as they could, by aid of rocks and bushes, and using their guns as staves. They reached the crest and swept it in a mighty fury. It was the decisive action. All the columns now converged on the distracted foe who fled before them. Grant galloped to the front with all speed, urging on the pursuit and exposing himself to every hazard of the fight.

So Chattanooga was added to Grant's lengthening score of brilliant victories; and again, as at Donelson and at Vicksburg, he had been the instrument of relieving a tense oppression of anxiety that had settled upon the nation. Sherman, with two corps, was at once sent to the relief of Knoxville; but Longstreet, having heard of Bragg's defeat, made an unsuccessful assault and retreated into Virginia. By the administration in Washington, and by the people of the North, General Grant's preëminence was conceded. His star shone brightest of all. Congress voted a gold medal for him.

CHAPTER XII

LIEUTENANT-GENERAL, COMMANDER OF ALL THE ARMIES

During the winter, after the Chattanooga victory, General Grant made his headquarters at Nashville, and devoted himself to acquiring an intimate knowledge of the condition of the large region now under his command, to the reorganization of his own lines of transportation, and the destruction of those of the enemy. He made a perilous journey to Knoxville in the dead of winter, and a brief trip to St. Louis, on account of the dangerous illness of his son there. On this trip he wore citizen's clothes, traveled as quietly as possible, declined all public honors, and made no delays. The whole route might have been a continuous enthusiastic ovation; but he would not have it so. His work was not done, and he sternly discountenanced all premature glorification. Too many generals had fallen from a high estate in the popular judgment, for him to court a similar fate. The promotions that gave him greater opportunity of service he accepted; but he preferred to keep his capital of popularity, whatever it might be, on deposit and accumulating while he stuck to his unaccomplished task, instead of drawing upon it as he went along for purposes of vanity and display. Of vulgar vanity he had as little as any soldier in the army.

Nashville was the base of supplies for all the operations in his military division. Its lines of transportation had been worn out and broken down, largely through incompetent management. He put them in charge of new men, who reconstructed and equipped them. While engaged in this necessary work he dispatched Sherman on an expedition through Mississippi, which he hoped would reach Mobile; but it terminated at Meriden, through failure of a cavalry force to join it. But it did a work in destruction of railroads and railroad property, that inflicted immense damage on the Confederacy.

Throughout the winter Grant worked as if his reputation was yet to be made, and to be made in that military division.

Meanwhile Congress and the country were pondering his deserts, and his ability for still greater responsibilities. The result of this deliberation was the passage of the act, approved March 1, 1864, reëstablishing the grade of lieutenant-general in the regular army. The next day President Lincoln nominated General Grant to the rank, and the nomination was promptly confirmed. He was ordered to Washington to receive the supreme commission. It was his first visit to the national capital; his first personal introduction to the President, although he had heard him make a speech many years before; his first meeting with the leading men in civil official life, who were sustaining the armies and guiding the nation in its imperiled way. He came crowned with the glory of victories second in magnitude and significance to none, since Wellington defeated Napoleon at Waterloo. Everybody desired to see him, and to honor him.

Yet he journeyed to Washington as simply and quietly as possible, avoiding demonstration. He arrived on the 8th of March, and going to a hotel waited, unrecognized, until the throng of travelers had registered, and then wrote, simply, "U. S. Grant and son, Galena." The next day, at 1 o'clock, he was received by President Lincoln in the cabinet-room of the White House. There were present, by the President's invitation, the members of the cabinet, General Halleck, and a few other distinguished men. After introductions the President addressed him as follows: —

"General Grant, — The expression of the nation's approbation of what you have already done, and its reliance on you for what remains to be done in the existing great struggle, are now presented with this commission, constituting you lieutenant-general in the army of the United States. With the high honor, devolves on you an additional responsibility. As the country herein trusts you, so, under God, it will sustain you. I scarcely need to add, that with what I here speak for the nation goes my own hearty personal concurrence."

General Grant made the following reply:—

"MR. PRESIDENT,—I accept the commission with gratitude for the high honor conferred. With the aid of the noble armies that have fought on so many battlefields for our common country, it will be my earnest endeavor not to disappoint your expectations. I feel the full weight of the responsibilities now devolving upon me; and I know that if they are met, it will be due to those armies; and, above all, to the favor of that Providence which leads both nations and men."

The next day he was assigned to the command of all the armies, with headquarters in the field. He made a hurried trip to Culpeper Court House for a conference with General Meade, commanding the Army of the Potomac; but would not linger in Washington to be praised and fêted. He hastened back to Nashville, where, on the 17th, he issued an order assuming command of the armies of the United States, announcing that until further notice, his headquarters would be with the Army of the Potomac. General Halleck was relieved from duty as general-in-chief; but was assigned by Grant to duty in Washington, as chief-of-staff of the army. Sherman was assigned to command the military division of the Mississippi, which was enlarged, and McPherson took Sherman's place as commander of the Army of the Tennessee; Thomas remaining in command of the Army of the Cumberland. On the 23d Grant was again in Washington, accompanied by his family and his personal staff. On the next day he took actual command, and immediately reorganized the Army of the Potomac in three corps, — the Second, Fifth, and Sixth, — commanded by Major-Generals Hancock, Warren, and Sedgwick; Major-General Meade retaining the supreme command. The cavalry was consolidated into a corps under Sheridan. Burnside commanded the Ninth Corps, which for a brief time acted independently.

This crisis of Grant's life should not be

passed over without allusion to the remarkable letters that passed between Grant and Sherman before he left Nashville to receive his new commission. Grant wrote to Sherman as follows: —

"Whilst I have been eminently successful in this war, in at least gaining the confidence of the public, no one feels more than I do how much of this success is due to the energy, skill, and the harmonious putting forth of that energy and skill, of those whom it has been my good fortune to have occupying subordinate positions under me. There are many officers to whom these remarks are applicable to a greater or less degree, proportionate to their ability as soldiers; but what I want is to express my thanks to you and McPherson as the men to whom, above all others, I feel indebted for whatever I have had of success. How far your advice and assistance have been of help to me, you know; how far your execution of whatever has been given you to do entitles you to the reward I am receiving you cannot know as well as I. I feel all the gratitude this letter would express, giving it the most flattering construction."

Grant's modesty, generosity, and magnanimity shine in this acknowledgment. If there were no other record illustrating these qualities, this alone would be an irrefragable testimony to his possession of them. There can be no appeal from its transparent, cordial sincerity.

Sherman's reply is too long to be quoted fully, but the parts of it that reveal his estimate of Grant's qualities and his confidence in him are important with reference to the purpose of this sketch: —

"You do yourself injustice and us too much honor in assigning to us too large a share of the merits which have led to your high advancement. . . . You are now Washington's legitimate successor, and occupy a position of almost dangerous elevation; but if you can continue, as heretofore, to be yourself, simple, honest, and unpretending, you will enjoy through life the respect and love of friends, and the homage of millions of human beings that will award you a large share in securing to them and their descendants a government of law and stability. . . . I believe you are as brave, patriotic, and just as the great prototype, Washington, as unselfish, kind-hearted, and honest as a man should be; but the chief characteristic is the simple faith in success you have always manifested, which I can liken to nothing else than the faith the Christian has in his Saviour. This faith gave you victory at Shiloh and Vicksburg. Also, when you have completed your preparations, you go into battle without hesitation, as at Chattanooga, — no doubts, no answers, — and I tell you it was this that made us act with confidence. I knew, wherever I was, that you thought of me; and if I got in a tight place you would help me out if alive."

He besought Grant not to stay in Washington, but to come back to the Mississippi Valley, "the seat of coming empire, and from the West where [when?] our task is done, we will make short work of Charleston and Richmond and the impoverished coast of the Atlantic." But Grant was wiser. He felt that the duty to which his new commission called him was to try conclusions with General Lee, the most illustrious and successful of the Confederate commanders, whom he had not yet encountered and vanquished. His new rank gave him an authority and prestige which would enable him, he trusted, to overcome the discouragements and discontents of the noble Army of the Potomac, and wield its unified force with victorious might. He knew, moreover, that the government and the people trusted him and would sustain him, as they trusted and would sustain no other, in a fresh and final attempt to destroy the Army of Northern Virginia, upon which the hopes of the Confederacy were staked. Not so much ambition as duty determined him to make his headquarters with the Army of the Potomac.

CHAPTER XIII

THE WILDERNESS AND SPOTTSYLVANIA

Wherever Grant had control in the West, and in all his counsels, his distinct purpose was to mass the Union forces and not scatter them, and to get at the enemy. With what ideas and intention he began the new task he set forth definitely in his report made in July, 1865.

"From an early period in the rebellion, I had been impressed with the idea that the active and continuous operations of all the troops that could be brought into the field, regardless of season and weather, were necessary to a speedy termination of the war. . . . I therefore determined, first, to use the greatest number of troops practicable against the armed force of the enemy, preventing him from using the same force at different seasons against first one and then another of our armies, and the possibility of repose for refitting and producing necessary supplies for carrying on resistance; second, to hammer continuously against the armed force of the enemy and his resources, until by mere attrition, if in no other way, there should be nothing left to him but an equal submission with the loyal sections of our common country to the Constitution and laws of the land."

Grant instructed General Butler, who had a large army at Fortress Monroe, to make Richmond his objective point. He instructed General Meade, commanding the Army of the Potomac, that Lee's army "would be his objective point, and wherever Lee went he would go also." He hoped to defeat and capture Lee, or to drive him back on Richmond, following close and establishing a connection with Butler's army there, if Butler had succeeded in advancing so far. Sherman was to move against Johnston's army, and Sigel, with a strong force, was to protect West Virginia and Pennsylvania from incursions. This, with plans for keeping all the other armies of the

Confederacy so occupied that Lee could not draw from them, constituted the grand strategy of the campaign.

The theatre of operations of the Army of the Potomac was a region of country lying west of a nearly north-and-south line passing through Richmond and Washington. It was about 120 miles long, from the Potomac on the north to the James on the south, and from 30 to 60 miles wide, intersected by several rivers flowing into Chesapeake Bay. The headquarters of the Union army were at Culpeper Court House, about 70 miles southwest of Washington, with which it was connected by railroad. This was the starting point. Lee's army was about fifteen miles away, with the Rapidan, a river difficult of passage, in front of it, the foothills of the Blue Ridge on its left, and on its right a densely wooded tract of scrub pines and various low growths, almost pathless, known as "the Wilderness."

Two courses were open to Grant, — to march by the right, cross the upper fords, and attack Lee on his left flank, or march by the left, crossing the lower fords, and making into the Wilderness. Grant chose the latter way, as, on the whole, most favorable to keeping open communications. For General Grant, as commander of all the armies, was bound to avoid being shut up or leaving Washington imperiled. And it may properly be said here that his plan contemplated leaving General Meade free in his tactics, giving him only general directions regarding what he desired to have accomplished, the actual fighting to be done under Meade's orders.

The official reports to the Adjutant-General's office in Washington show that on the 20th of April the Army of the Potomac numbered 81,864 men present and fit for duty. Burnside's corps, which joined in the Wilderness, added to this force 19,250 men, making a total of 101,114 men. After the Wilderness, a division numbering 7000 or 8000 men under General Tyler joined it. When the Chickahominy was reached, a junction with Butler's army, 25,000 strong, was made. Lee had on the 20th of April present for duty, armed and equipped, 53,891. A few days later he was reinforced by Longstreet's corps, which on the date given numbered 18,387, making a total of 72,278. Grant's army outnumbered Lee's, but he was to make an offensive campaign in the enemy's country, operating on exterior lines, and keeping long lines of communication open. Defending Richmond and Petersburg there were other Confederate forces, under Beauregard, Hill, and Hoke, estimated to amount to nearly 30,000 men, and Breckenridge commanded still another army in the Shenandoah Valley. In Grant's command, but not of the Army of the Potomac, were the garrison of Washington and the force in West Virginia.

On the 3d of May the order to move was given, and at midnight the start was made. The advance guard crossed the river before four in the morning of the 4th, and on the morning of the 5th Grant's army, nearly a hundred thousand strong, was disposed in the Wilderness. Lee had discovered the movement promptly, and had moved his whole army to the right, determined to fall upon Grant in that unfavorable place. As soon as the Union army began a movement in the morning, it encountered the enemy, who attacked with tremendous and confident vigor. The fighting continued all day, with indecisive results. Early the next morning the battle was renewed, and continued with varying fortunes, at one time one army, and at another time the opposing army, having the advantage. There was, in fact, a series of desperate battles between different portions of the two armies which did not end until the night was far advanced. The advantage, on the whole, was with the Union army. It had not been forced back over the Rapidan. It stood fast. But it had inflicted no such defeat on the enemy as Grant had hoped to do in the first encounter. The losses of both sides had been very large, those of the Union Army being 3288 killed, 19,278 wounded, 6784 missing.

The next morning it was discovered that the Confederates had retired to their intrenchments, and were not seeking battle.

Then Grant gave the order that was decisive, and revealed to the Army of the Potomac that it had a new spirit over it. The order was, "Forward to Spottsylvania!" No more turning back, no more resting on a doubtful result. "Forward!" to the finish. But Lee, controlling shorter lines, was at Spottsylvania beforehand, and had seized the roads and fortified himself. Here again was bloody fighting of a most determined character, lasting several days. Here Hancock, by a daring assault, captured an angle of the enemy's works, with a large number of guns and prisoners; and it was held, despite the repeated endeavors of the enemy to recapture it. Here General Sedgwick was killed. Here Upton made a famous assault on the enemy's line and broke through it, want of timely and vigorous support preventing this exploit from making an end of Lee's army then and there. But the Union losses at Spottsylvania, while not so large as in the Wilderness, were very heavy, and made a painful impression upon the people of the North.

Undoubtedly Grant was disappointed by the failure to vanquish his opponent. Undoubtedly Lee was disappointed by his failure to repulse the Union army in the Wilderness and at Spottsylvania as he had done formerly at Chancellorsville and Fredericksburg, when it had come into the same territory. Each had underestimated the other's quality. From Spottsylvania, on the 11th of May, after six days of continuous fighting, with an advance of scarcely a dozen miles, and an experience of checks and losses that would have disheartened any one but the hero of Vicksburg, he sent this bulletin to the War Department: "We have now ended the sixth day of very heavy fighting. The result to this time is much in our favor. But our losses have been heavy, as well as those of the enemy. We have lost to this time 11 general officers killed, wounded, and missing, and probably 20,000 men. . . . I am now sending back to Belle Plain all my wagons for a fresh supply of provisions and ammunition, and propose to fight it out on this line if it takes all summer."

The indomitable spirit of the last sentence electrified the country. It did take all summer, and all winter, too, — eleven full months from the date of this dispatch, and more, before General Lee, driven into Richmond, forced to evacuate the doomed city, his escape into the South cut off, his soldiers exhausted, ragged, starving, reinforcements out of the question, surrendered at Appomattox the Army of Northern Virginia, the reliance of the Confederacy, to the general whom he expected to defeat by his furious assault in the Wilderness.

CHAPTER XIV

FROM SPOTTSYLVANIA TO RICHMOND

THE story of this campaign is too long to be narrated in particular. On both sides it is a record of magnificent valor, endurance, and resolution, to which the world affords no parallel, when it is remembered that the armies were recruited from the free citizenship of the nation. As the weeks and months wore on, General's Grant's visage, it is said, settled into an unrelaxing expression of grim resolve. He carried the nation on his shoulders in those days. If he had wearied or yielded, hope might have vanished. He did not yield nor faint. He planned and toiled and fought, keeping his own counsel, bearing patiently the disappointment, the misunderstanding, the doubt, the criticism, the woe of millions who had no other hope but in his success and were often on the verge of despair. He beheld his plans defeated by the incompetence or errors of subordinates whom he trusted, and let the blame be laid upon himself without protest or murmuring. He knew better than any one else the terrible cost of life which his unrelenting purpose demanded; but he knew also that the price of relenting, involving the discouragement of failure, the cost of another campaign after the enemy had got breath and new equipment, the possible refusal of the North to try again, was far greater and more humiliating. Little wonder that he was oppressed and silent and moody. Yet he ruled his own spirit in accordance with the habit of his life. No folly or disappointment provoked him to utter an oath. General Horace Porter, of his staff, a member of his intimate military family, says that the strongest expression of vexation that ever escaped his lips was: "Confound it!" He alone had the genius to be master of the situation at all times, and the "simple faith in success" that would not let him be swerved from his aim.

So he pressed on from the Wilderness to

Spottsylvania, to North Anna, to South Anna, to the Pamunky, to Cold Harbor, to the Chickahominy, fighting and flanking all the way, until at the end of the month he had pressed Lee back to the immediate vicinity of Richmond. The bloodiest of all these battles was the ill-judged attack, for which Grant has been much criticised, on the strongly intrenched rebel lines at Cold Harbor. If he could have dislodged Lee here he could have compelled him to retreat into the immediate fortifications of Richmond. But Lee's position was impregnable: the assault failed. In less than an hour Grant lost 13,000 men killed, wounded, and missing, and gained nothing substantial.

General Butler had signally failed to accomplish the work given him to do. Instead of taking Petersburg, destroying the railroads connecting Richmond with the south, and laying siege to that city, he had, after some ineffectual manœuvring, got his army hemmed in, "bottled up," Grant called it, at Bermuda Hundred, where he was almost completely out of the offensive movement for months. Sigel had been worsted in the North, and had been relieved by Hunter, who had won measurable success in the Shenandoah Valley.

Grant, checked on the east and north of Richmond, crossed the Chickahominy and the James with his whole army by a series of masterly manœuvres, regarding the meaning of which his opponent was brilliantly deceived. Then followed the unsuccessful attempt to capture Petersburg before it could be reinforced, unsuccessful by reason of the want of persistence on the part of the general intrusted with the duty. This failure involved a long siege of that place, which the Confederates made impregnable to assault. A breach in the defences was made by the explosion of a mine constructed with vast labor, but there was failure to follow up the advantage with sufficient promptness. Here the Army of the Potomac passed the winter, except the part of the army that was detached to protect Washington from threatened attack, and with which Sheridan made his great fame in the Shenandoah Valley. Meanwhile Sherman, in the West, had taken Atlanta, and leaving Hood's army to be taken care of by Thomas, who defeated it at Nashville, had marched across Georgia, and was making his way through the Carolinas northward toward Richmond, an army under Johnston disputing his way by annoyance, impediment, and occasional battle. Another incident of the winter was the two attempts on Fort Fisher, near Wilmington, North Carolina, — the first, under General Butler, a failure; the second, under General Terry, a brilliant success. All these movements were in execution of plans and directions given by the lieutenant-general.

It was the 29th of March when, all preparations having been made, Grant began the final movement. He threw a large part of his army into the region west of Petersburg and south of Richmond, and at Five Forks, four days later, Sheridan fought a brilliant and decisive battle, which compelled Lee to abandon both Petersburg and Richmond, and to attempt to save his army by running away and joining Johnston. All his movements were baffled by the eager Union generals, flushed with the consciousness that the end was near.

On the 7th of April Grant wrote to Lee: "I regard it as my duty to shift from myself responsibility for any further effusion of human blood by asking of you the surrender of that portion of the Confederate States army, known as the Army of Northern Virginia." Lee replied at once, asking the terms that would be offered on condition of surrender. His letter reached Grant on the 8th, who replied: "*Peace* being my great desire, there is but one condition I would insist upon, namely: that the men and officers surrendered shall be disqualified for taking up arms again against the government of the United States until properly exchanged." He offered to meet Lee or any officers deputed by him for arranging definite terms. Lee replied the same evening somewhat evasively, setting forth that he desired to treat for peace, and that the surrender of his army would be considered as a means to that end.

To this Grant responded on the 9th, having set his army in motion to Appomattox Court House, that he had no authority to treat for peace; but added some plain words to the effect that the shortest road to peace would be surrender. Lee immediately asked for an interview. Grant received this communication while on the road, and returned word that he would push on and meet him wherever he might designate. When Grant arrived at the village of Appomattox Court House he was directed to a small house where Lee awaited him. Within a short time the conditions were drafted by Grant and accepted by Lee, who was grateful that the officers were permitted to keep their side-arms, and officers and men to retain the horses which they owned and their private baggage.

The number of men surrendered at Appomattox was 27,416. During the ten days' previous fighting 22,079 of Lee's army had been captured, and about 12,000 killed and wounded. It is estimated that as many as 12,000 deserted on the road to Appomattox.

From May 1, 1864, to April 9, 1865, the Armies of the Potomac and the James took 66,512 prisoners and captured 245 flags, 251 guns, and 22,633 stands of small arms. Their losses from the Wilderness to Appomattox were 12,561 killed, 64,452 wounded, and 26,988 missing, an aggregate of 104,001.

It would be idle adulation to say that in all points during this long conflict with Lee General Grant always did the best thing, making no mistakes. The essential point is, and it suffices to establish his military fame on secure foundations, that he made no fatal mistake, that progress toward the great result in view was constant, slower than he expected, slower than the country expected, but finally everywhere victorious, substantially on the lines contemplated in the beginning. After Lee's experience in the Wilderness and at Spottsylvania he seldom assumed the offensive against Grant. He became prudent, adopted a defensive policy, fought behind intrenchments or just in front of fortifications to which he could retire for safety, and waited to be attacked. Watchful and alert as he was, he was deceived by Grant oftener than he deceived him, and except that he managed to postpone the end by skillful tactics, he did not challenge the military superiority of his foe. He made Grant's victory costly and difficult, but he did not prevent it. He retreated with desperate reluctance, but he was forced back. He could not protect his capital; he could not save his army. When Lee measured powers with Grant, his cause was lost.

There are incidents of the campaign that mitigate its stern and in some sense savage features. When the imperturbable soldier learned of the death of his dear friend McPherson, who fell in one of Sherman's battles, he retired to his tent and wept bitterly. When Lincoln, visiting Grant at City Point, before the general departed on what was expected to be the last stage of the campaign, said to him that he had expected he would order Sherman's army to reinforce the Army of the Potomac for the final struggle, the reply was that the Army of the Potomac had fought the Army of Virginia through four long years, and it would not be just to require it to share the honors of victory with any other army. It was observed that when he bade good-by to his wife at this departure his adieus, always affectionate, were especially tender and lingering, as if presentiment of a crisis in his life oppressed him. Lincoln accompanied him to the train. "The President," said Grant, after they had parted, "is one of the few who have not attempted to extract from me a knowledge of my movements, although he is the only one who has a right to know them." Long before, Lincoln had written to him: "The particulars of your campaign I neither know nor seek to know. I wish not to intrude any restraints or constraints upon you." Grant's reply to this confidence was: "Should my success be less than I desire or expect, the least I can say is, the fault is not yours." These two understood each other by a magnanimous sympathy that had no need of particular confidences. That Lincoln respected Grant as one whom it was not becoming for him to presume to question is in itself impressive evidence of Grant's greatness.

CHAPTER XV

IN WASHINGTON AMONG POLITICIANS

WITHIN a few weeks after the surrender of Lee, every army and fragment of an army opposed to the Union was dissolved. But meantime Lincoln had been assassinated, and the executive administration of the nation had devolved upon Andrew Johnson. This wrought an immense change in the aspect of national affairs. Lincoln was a strong, wise, conservative, magnanimous soul. Johnson was arrogant, vain, narrow, and contentious. Grant soon established his headquarters at the War Department, and devoted himself with characteristic energy to the work of discharging from the military service the great armies of volunteers no longer needed. Their work as soldiers was gloriously complete. Within a few months they were once more simple citizens of the Republic, following the ways of industry and peace. The suddenness of the transformation by which at the outbreak of hostilities hundreds of thousands of citizens left their homes and their occupations of peace to become willing soldiers of the Union and liberty, was paralleled by the alacrity of their return, the moment the danger was passed, to the stations and the manner of life they had abandoned.

General Grant was the central figure in the national rejoicing and pride. The desire to do him honor was universal. But he bore himself through all with dignity and modesty, avoiding as much as he could, without seeming inappreciation and disdain, the lavish popular applause that greeted him on every possible occasion. In July, 1866, Congress created the grade of general, to which he was at once promoted, thus attaining a rank never before granted to a soldier of the United States. His great lieutenant, Sherman, succeeded him in this office, which was then permitted to lapse, though it was revived later as a special honor for General Sheridan. In further token of gratitude, some of the wealthier citizens purchased and presented to Grant a house in Washington. Resolutions of gratitude, honorary degrees, presents of value and significance, came to him in abundance. Through it all, he maintained his reputation as a man of few words, devoid of ostentation, and with no ambition to court public favor by any act of demagoguism.

But a great and bitter trial confronted him. He had never been a politician. Now he was caught in a maelstrom of ungenerous and malignant politics. All his influence and effort had been addressed to promote the calming of the passions of the war, and a reunion in fact as well as in form. The President, professing an intention of carrying out the policy of his predecessor, began a method of reconstructing civil governments in the States that had seceded which produced great dissatisfaction. Upon his own initiative, without authority of Congress, he proceeded to encourage and abet those who were lately in arms against the Union to make new constitutions for their States, and institute civil governments therein, as if they alone were to be considered. The freedmen, who had been of so great service to our armies, whom by every requirement of honor and gratitude we were bound to protect, were left to the hardly restricted guardianship of their former masters, who, having no faith in their manhood or their development, devised for them a condition with few rights or hopes, and little removed from the slavery out of which they had been delivered.

This policy found little favor with those in the North who had borne the heat and burden of the war. In the elections of 1866 the people repudiated President Johnson's policy by emphatic majorities. When the hostile Congress met, the governments Johnson had instituted were declared to be provisional only, and it set about the work of reconstruction in its own way, imbedding the changed conditions, the fruits of the war, in proposed amendments of the Constitution of the United States, which were ultimately ratified by a sufficient number of States to make them part of the organic frame of government of the Republic.

In these days of storm and stress, General Grant took neither side as a partisan. He stuck to his professional work until he was forced to be a participator in a political war, strange to his knowledge and his habits. Congress directed the Southern States to be divided into five military districts, with a military commander of each, and all subordinate to the general of the army, who was charged with keeping the peace, until civil governments in the States should be established by the legislative department of national authority.

Congress, before adjourning in 1866, passed a tenure-of-office act, — overriding in this, as in other legislation, the President's veto. The motive was to prevent the President from using the patronage to strengthen his policy. This act required the President to make report to the Senate of all removals during the recess, with his reasons therefor. All appointments to vacancies so created were to be *ad interim* appointments. If the Senate disapproved of the removals, the officer suspended at once became again the incumbent. Severe penalties were provided for infraction of the law. During the recess the President removed Stanton, and appointed General Grant to be Secretary of War. Grant did not desire the office, but under advice accepted it, lest a worse thing for the country might happen.

Johnson hoped to win Grant to his side, and in any event to use him in his strife with Congress to defeat the purpose of the law. While the Senate had Stanton's case under consideration in January, 1867, Grant was called into a cabinet meeting and questioned regarding what he would do. He said that he was not familiar with the law, but would examine it and notify the President. The next day he notified him that he would obey the law. Therefore, when the Senate disapproved of the reasons assigned for the removal of Stanton, Grant at once vacated the office, to the intense mortification and anger of the President, who made a public accusation that Grant had promised to stay in office and oppose Stanton's resumption of it.

The charge made a great scandal, but it did not seriously impair Grant's good repute. Johnson was not believed, and the testimony of the members of his cabinet, regarding what happened, was so conflicting that it failed to convince anybody who did not seek to be convinced.

There is reason to believe that Johnson never contemplated retaining Grant in the office, except to use his name and fame to break down the tenure-of-office act. General Grant's plain common sense delivered him from the snare spread for him by wily and desperate politicians. On February 3, he closed an unsatisfactory correspondence with President Johnson, with these severe words: "I can but regard this whole business, from the beginning to the end, as an attempt to involve me in the resistance of law, for which you hesitated to assume the responsibility in orders, and thus to destroy my character before the country. I am, in a measure, confirmed in this conclusion by your recent order, directing me to disobey orders from the Secretary of War, my superior and your subordinate, without having countermanded his authority to issue the orders I am to disobey."

When Johnson was impeached by the House of Representatives, General Grant might, if he had chosen to do so, have contributed much to embarrass the President; but he held aloof, discharging his duties as general-in-chief with constant devotion. He was instrumental in instituting many economies and improvements of army management. He greatly advanced the work of reconstruction, and civil governments were firmly established on the congressional plan in a majority of the Southern States before he became the chosen leader of the Republican party.

Grant had not yet distinctly committed himself as between the Democratic and the Republican parties, although from the time of his break with Johnson, he was more drawn to the Republicans. So far as he had any politics he might have been classed as a War Democrat. Had he definitely proclaimed himself a Democrat, no doubt he could have had that party's nomination for the presidency. He was the first citizen of the nation in popularity, of which he had marked tokens, and of which both parties were anxious to avail themselves. It is little wonder that he came to think that the presidency was an honor to which he might fitly aspire, and an office in which he could further serve his country, by promoting good feeling between the sections. In May, 1868, he was placed in nomination, first by a convention of Union soldiers and sailors, and afterwards by the Republican party, in both instances by acclamation. His Democratic opponent was Horatio Seymour, of New York. In the election he had a popular majority of 305,456. He received 214 electoral votes, and Seymour received 80. Three of the Southern States, not being fully restored to the Union, had no voice in the election.

CHAPTER XVI

HIS FIRST ADMINISTRATION

IMMEDIATELY after General Grant's inauguration as President, an incident occurred which revealed his inexperience in statesmanship. Among the names sent to the Senate as members of the cabinet was that of Alexander T. Stewart, of New York, the leading merchant of the country, for Secretary of the Treasury. Grant was unaware of the existence of his disqualification by a statute passed in 1789, on account of being engaged in trade and commerce. His ignorance is hardly surprising in view of the fact that the Senate confirmed the nomination without discovering its illegality. The point was soon made, however, and the reasonableness of the law was apparent to all except the President, who sent a message to the Senate suggesting that Mr. Stewart be exempted from its application to him by a joint resolution of Congress. This breaking down of a sound principle of government for the pleasure of the President was not favored, and George S. Boutwell of Massachusetts was substituted, Mr. Stewart having declined, in order to relieve the President of embarrassment.

For the rest, the cabinet was a peculiar one. It appeared to be made up without consultation or political sagacity, in accordance with the personal reasons by which a general selects his staff. Elihu B. Washburn, of Illinois, his firm congressional friend during the war, was Secretary of State; General Jacob D. Cox, of Ohio, Secretary of the Interior; Adolph E. Boise, of Pennsylvania, Secretary of the Navy; General John M. Schofield, of Illinois, Secretary of War; John A. J. Cresswell, of Maryland, Postmaster-General; and E. Rockwood Hoar, of Massachusetts, Attorney-General. It did not long endure in this form. Mr. Washburn was soon appointed Minister to France, and was succeeded by Hamilton Fish, of New York, in the State

Department. General Schofield was succeeded in the War Department by General John A. Rawlins, who died in September, and was succeeded by General William W. Belknap, of Iowa. Mr. Boise gave way in June to George M. Robeson, of New Jersey. In July, 1870, Mr. Hoar was succeeded by A. T. Akerman, of Georgia, and he, in December, 1871, by George H. Williams, of Oregon. General Cox resigned in November, 1870, and was succeeded by Columbus Delano. Some of these changes, like that of Washburn to Fish, were good ones, and many of them were exceedingly bad ones, —men of high character and ability, like Judge Hoar and General Cox, conscientious and faithful even to the point of remonstrance with their headstrong chief, being succeeded by compliant men of a distinctly lower strain. Fish and Boutwell achieved high reputation by their conduct of their offices. The death of Rawlins deprived the President of a wise and staunch personal friend at a time when he was never more in need of his controlling influence.

Early in 1871 the work of reconstruction was completed, so far as the establishment of State governments and representation in Congress was concerned. But later in the year, the outrages upon the colored population in certain States were so general and cruel that Congress passed what became known as the "Ku-Klux Act," which was followed by a presidential proclamation exhorting to obedience of the law. On October 17, the outrages continuing, suspension of the writ of habeas corpus was proclaimed in certain counties of South Carolina, and many offenders were convicted in the United States courts. This severe proceeding had a deterring influence throughout the South, which understood quite well that General Grant was not a person to be defied with impunity.

In 1870 he sent to the Senate a treaty that the administration had negotiated with President Baez for the annexation of Santo Domingo as a territory of the United States, and also one for leasing to the United States the peninsula and bay of Samana. These treaties, it was said, had already been ratified by a popular vote early in 1870. The scheme precipitated a conflict that divided the Republican party into administration and anti-administration factions, the latter being led by Charles Sumner and Carl Schurz. Sumner had long been chairman of the Senate committee on foreign relations, but he was degraded through the influence of the President's friends in the Senate. Bitter personal animosities were aroused in this contest which never were healed. It was alleged that the sentiment of the people of Santo Domingo had not been fairly taken, and that they were in fact opposed to annexation. A commission composed of B. F. Wade, of Ohio, Andrew D. White, of New York, and Samuel G. Howe, of Massachusetts, was sent on a naval vessel to investigate the actual conditions. This committee reported in favor of annexation; but the hostile sentiment in Congress and among the people was so strong that the treaties were never ratified. By many it was considered a wrong to the colored race to so extinguish the experiment of negro self-government. Others were opposed to annexing such a population, thinking this country already had race troubles enough. Others regarded the whole business as a speculation of jobbers, and the stain of jobbery then pervading government circles was so notorious that the presumption was not without warrant. The annexation scheme brought to a head and gave occasion for an outbreak of indignant hostile criticism of the President and the administration.

In this term Grant appointed the first board of civil service commissioners, with George William Curtis at its head. The commissioners were to inquire into the condition of the civil service and devise a scheme to increase its efficiency. This they did; but later the President himself balked at the enforcement of their rules, and, in 1873, Mr. Curtis resigned.

The most conspicuous achievement of General Grant's first term was the settlement of the controversy with Great Britain growing out of the destruction of American commerce by Confederate States cruisers during the war. A joint high commission of five British and five American members met in Washington, February 17, 1871, and on May 8 a treaty was completed and signed, providing peaceable means for a settlement of the several questions arising out of the coast fisheries, the northwestern boundary line, and the "Alabama Claims." The last and most important subject was referred to an international court of arbitration, which met at Geneva, Switzerland, and on September 14, 1872, awarded to the United States a gross sum of $15,500,000, which was paid by Great Britain. This was the most important international issue that had ever been settled by voluntary submission to arbitration. It was long regarded as the harbinger of peace between nations.

Other important things done were the establishment of the first weather bureau; the honorable settlement of the outrage of Spain in the case of the Virginius, an alleged filibustering vessel which Spain seized, executing a large part of its crew in Cuba;

and the settlement of the northwest boundary question. It should be said also that the President made a firm stand in behalf of national financial integrity.

But during the four years there was a steady deterioration in the tone of official life, and a steady growth of corruption and abuses in the administration of government. The President exhibited a strange lack of moral perception and stamina in the sphere of politics. Unprincipled flatterers, adventurers, and speculators gained a surprising influence with him. His native obstinacy showed itself especially in insistence upon his personal, ill-instructed will. He became intractable to counsels of wisdom, and seemed to be a radically different man from the sincere, modest soldier of the civil war. He affected the society of the rich, whom he never before had opportunity of knowing. He accepted with an indiscreet eagerness presents and particular favors from persons of whose motives he should have been suspicious. Jay Gould and James Fisk used him in preparing the conditions for the corner of the gold market that culminated in "Black Friday." He provided fat offices for his relatives with a liberal hand, and prostituted the civil service to accomplish his aims and reward his supporters.

In consequence of these things there was great disaffection in the Republican party, which culminated in open revolt. Yet he was supported by the majority. The Democratic party, meantime, making a virtue of necessity, proclaimed a purpose to accept the results of the war, including the constitutional amendments, as accomplished facts not to be disturbed or further opposed. This made an opportunity for a union of all elements opposed to the reëlection of Grant, leading Democrats having given assurance of support to a candidate to be nominated by what had come to be called the "Liberal Reform" party. That party held its convention in Cincinnati early in May, and named Horace Greeley as its candidate, a nomination which wrecked whatever chance the party had seemed to have. Grant was renominated by acclamation in the Republican convention. The Democratic convention nominated Greeley on the Cincinnati convention platform, but without enthusiasm. General Grant was elected by a popular majority of more than three quarters of a million, and a vote in the electoral college of 286 to 63 for all others, the opposing vote being scattered on account of the death of Mr. Greeley in November, soon after his mortifying defeat.

CHAPTER XVII

HIS SECOND ADMINISTRATION

The storm of criticism and calumny through which President Grant passed during the election canvass of 1872 had no effect to change his general course or open his eyes to the true sentiment of the nation. Instead of realizing that he was reëlected, not because his administration was approved, but because circumstances prevented an effective combination of the various elements of sincere opposition, he and his friends accepted the result as popular approbation of their past conduct and warrant for its continuance. Things went from bad to worse with a pell-mell rapidity that made good men shudder.

In the four years there were but two exhibitions of conspicuously courageous and honorable statesmanship. One was the passage of the Resumption Act of January 14, 1875, which promised the resumption of specie payments on January 1, 1879, and gave the Secretary of the Treasury adequate power to make the performance of the promise possible. This was one result of the collapse in 1873 of the enormous speculation promoted by a fluctuating currency and fictitious values. The demand for a currency of stable value enabled the conservative statesmen in Congress to take this action. Grant's approval of this act and his veto in the previous year of the "inflation bill" must always be regarded as highly commendable public services.

The only immediate change in the cabinet was the appointment of William A. Richardson to succeed George S. Boutwell as Secretary of the Treasury. Mr. Richardson had some qualifications of experience for the place, but wanted the essential traits of firmness and high motive. In the next year after taking office he was forced to resign, on account of a report of the committee of ways and means condemning him for his part in making a contract, while acting

Secretary of the Treasury, with one Sanborn, for collecting for the Treasury, on shares, taxes which it was the business of regular officers of the government to collect. Immense power was given by the contract, and the resources of the Treasury Department were put at the service of a crew of irresponsible inquisitors before whom the business community trembled. They extorted immense sums in dishonorable ways which aroused popular resentment. The President saw no wrong, and accepted Secretary Richardson's resignation unwillingly, at once nominating him to be Chief Justice of the Court of Claims, a reward for malfeasance which amazed the country, although the administration supporters in the Senate confirmed it.

General Benjamin H. Bristow, of Kentucky, became Secretary of the Treasury, a man of superior ability, aggressive honesty, and moral firmness. He quickly uncovered a mass of various wrongdoing, — the safe-burglary frauds of the corrupt ring governing Washington, the seal-lock frauds, the subsidy frauds, and, most formidable of all, the frauds of the powerful whiskey ring having headquarters in St. Louis. The administration of the Treasury Department, especially the Internal Revenue Bureau, was permeated with corruption. The worst feature of it all was that officers who desired to be upright found themselves powerless against the intrigues and the potent political influence of the rascals at the headquarters of executive authority. When the evidence of wrongdoing accumulated by the new Secretary of the Treasury was laid before the President he was dumfounded by its wickedness and extent, but showed himself resolute and vigorous in supporting his able and resourceful Secretary. The trap was sprung in May, 1875. Indictments were found against 150 private citizens and 86 government officers, among the latter the chief clerk in the Treasury Department, and the President's private secretary, General O. E. Babcock. All the principal defendants were convicted except Babcock, and he was dismissed by the President.

During all these proceedings, in spite of the President's professions, the Treasury Department was beset by subtle hostile influences and impediments. The politicians who had the President's ear made him believe that it was the ruin of himself and his household that the investigators sought. Only the enthusiastic popular approval of Secretary Bristow's brave course prevented yielding to the political backers of the corruption. When in the spring of 1876 Bristow initiated a similar campaign against the corruptions rife on the Pacific coast, the Secretary was overruled and the government prosecutors were recalled. Whereupon the Secretary resigned, and no less than seven high Treasury officials, who had been active in the crusade of reform, left the department at the same time. Mr. Bristow was succeeded by an honorable man, — the President had to appoint a man known to be pure, — Lot M. Morrill, of Maine; but he was infirm, and all aggressive reform work ceased.

In the War Department, Secretary Belknap, sustained by the President, stripped General Sherman of the rights and duties properly pertaining to his rank, of which Grant himself, in the same place during Johnson's administration, had protested against being deprived. Sherman was subjected to such humiliations by his old commander, turned politician, that he abandoned Washington and retired to St. Louis. Congress was a subservient participator in this shame, repealing the law that required all orders to the army to go through its general. But in February, 1876, it was discovered that Belknap had been enriching himself by corrupt partnership with contractors in his department, and he hurriedly resigned, the President strangely accepting the resignation before Congress could act. He was impeached, notwithstanding. He set up the defense that being no longer an official, he could not be impeached, and this being overruled, he was tried, but was not convicted. Of his guilt the country had no doubt. Then Alphonso Taft, an Ohio judge, was made Secretary of War. He was soon transferred to the Attorney-General's office, and was succeeded by Don Cameron, already his father's lieutenant in control of the Republican party of Pennsylvania.

Columbus Delano, Secretary of the Interior, had so mismanaged affairs, especially in the Indian Bureau, which teemed with flagrant abuses, that public opinion turned against him with great force, and in 1875 he had to abandon the office, in which he was succeeded by Zachariah Chandler, against whom no scandalous charge was made, although he was a rank partisan of the President.

Marshall Jewell, of Connecticut, became Postmaster-General in 1874. He was a successful business man, and on taking the office he declared his purpose to conduct it on business principles. He attacked effectively a system long in vogue known as "straw-bids" for mail-carrying contracts. He introduced the railway post-office system, that has been of so much use in facilitating promptness of transmitting correspondence. But he also insisted on conducting his office with respect of its personnel as a business man would, that is, by making appointments and promotions for merit rather than for political influence. This was intolerable to the spoilsmen in politics; and within two years he was summarily dismissed in a manner as graceless and cruel as any President, no matter how unfortunately bred, was ever guilty of. Jewell was succeeded by James N. Tyner, an entirely complaisant official. In 1875 Congress neglected to make any appropriation for the civil service reform commission, and its work was suspended.

During this time affairs in the Southern States were, as a rule, growing worse and worse. The unreasonable arrogance and oppressive extravagance of the freedmen where they were in control, under the leadership of reckless carpet-baggers, and still more reckless and malicious white natives, had produced a revulsion in the minds of all at the North who regarded justice, honor, and honesty as essentials of good government. There were exceptions, like oases in the desert of ignorance and vice. The administration of Governor Chamberlain in South Carolina was an instance of an earnest and partially successful endeavor to educe good government from desperate conditions. The colored race abused its privilege of the ballot with suicidal persistency. The experiment of maintaining bad State governments by the presence and activity of federal troops did not tend to social pacification. Reconstruction in its earlier fruits was an obvious failure; and again, if the apparent paradox can be understood, lawless violence began asserting itself as the only hopeful means of preserving property, civil rights, and civilization itself.

During the second term the report was persistently circulated that Grant and those who followed his star were scheming for another term, in order to give him in civil office, as in military rank, a distinction higher than Washington or any American had obtained. The proposal shocked the public sense of propriety; but its treatment by those who alone could repudiate it became ominous. The Republican State Convention of 1875 in Pennsylvania boldly declared unalterable opposition to the third-term idea. Grant then spoke. In a letter to the convention's chairman he said: "Now, for the third term, I do not want it any more than I did the first." After calling attention to the fact that the Constitution did not forbid a third term, and that an occasion might arise when a third term might be wisely given, he said that he was not a candidate for a third nomination, and "would not accept it, if tendered, unless under such circumstances as to make it an imperative duty — circumstances not likely to arise."

This was justly regarded as a politician's letter, and increased alarm instead of allaying it. The national House of Representatives (which the elections of 1874 had made a Democratic body), by a vote of 234 to 18, passed the following resolution: "That in the opinion of this House the precedent, established by Washington and other Presidents of the United States after their second term, has become, by universal consent, a part of our republican system of government, and that any departure from this time-honored custom would be unwise, unpatriotic, and fraught with peril to our free institutions." As 70 Republicans voted for this resolution, it was practically the voice of both parties, and it dispelled the spectre of "Cæsarism," as the third-term idea was called. There is reason to believe that if it had caused less alarm it would have assumed a more substantial aspect.

During the excited and perilous four months after the election of 1876, when civil war and anarchy were imminent on account of the disputed result of the people's suffrage, the conduct of the President was admirable. He let it be understood that violence would be suppressed, without hesitation, at any cost. He preserved the *status quo*, and compelled peaceful patience. The condition was one which summoned into action his genius of supreme command, and it shone with its former splendor of authority. On the 4th of March, 1877, he became a private citizen.

CHAPTER XVIII

THE TOUR OF THE WORLD

UPON leaving the presidency General Grant retained the distinction of first citizen of the nation. There was no fame of living man that could vie with his. His old form of modesty and simplicity was resumed. As soon as he stepped down from the pedestal of power the criticism of duty and the criticism of malice both ceased. A generous people was glad to forget his errors and remember only his patriotism and his transcendent successes in arms. Even those who had most deprecated his mistakes as a civil magistrate were hardly sorry that he had been repeatedly rewarded for his great services by the highest honor popular suffrage could bestow. They were ready to believe, as, indeed, was true, that in most of the things deserving reprobation he was the victim of his innocence of selfish politics and his unwary friendships, of which baser men had taken foul advantage. They were glad for his sake, as much as for their own, that he was no longer President Grant, but again General Grant, a title purely reminiscent and complimentary, for he was no longer an officer of the army. With all his honors about him, he stood on the common level of citizenship, as when he was a farmer in Missouri or a tanner's clerk in Galena.

There came to him then the desire to see other lands and peoples and to meet the renowned commanders in other wars, the actors in other statesmanship. It was determined that he should have all the opportunities and advantages which the national prestige could command for its foremost unofficial representative. No other American had gone abroad whose achievements bespoke for him so respectful a welcome among the great. Every aid was availed of to make it apparent that our nation expected him to be entertained as its beloved hero. He sailed from Philadelphia on May 17, 1877, and, returning, he landed in San

Francisco September 20, 1879, having made the circuit of the globe.

Of such another progress there is no record. He visited nearly every country of Europe, the Holy Land, Egypt, Syria, India, Burmah, China, Siam, and Japan, being everywhere received as the guest of their rulers, and welcomed by the chief representatives of their statesmanship, their learning, and their social life. He was received with high courtesies by Queen Victoria of England, President McMahon and President Grévy of France, the emperors of Germany, Russia, and Austria, the kings of Belgium, Italy, Holland, Sweden, and Spain, Pope Leo XIII., the Sultan of Turkey, the Khedive of Egypt, the Duke of Wellington, Prince Bismarck, M. Gambetta, Lord Lytton, Viceroy of India, King Thebau of Burmah, Prince Kung of China, the Emperor of Siam, the Mikado of Japan, and many others only less famous. With few exceptions he met under the most favorable circumstances all persons of note in all the lands he visited. Extraordinary pains were taken to promote the comfort of his party, and to enable its members to see whatever was most worth seeing.

The recipient of all this flattering attention bore himself with a simple dignity that won the respect of the high and the low alike. He was neither awed nor abashed among the great, nor was he haughty or presuming among the common people. The nation at home followed his progress with pride and gratification. When he landed in San Francisco, he was welcomed as a favorite who had achieved new distinction for himself and his land, and his leisurely way across the continent was marked by a series of ovations all the way to New York. To complete his itinerary, he soon made a tour of the West Indies and of Mexico, visiting the scenes where he had won his first laurels, as Lieutenant Grant, thirty years before. He was honored as the warrior whose victories, besides uniting and exalting his native land, had delivered Mexico from the imposition of an alien imperialism.

Unfortunately, this revived popularity of General Grant was taken advantage of by a faction of the Republican party to urge again his reëlection to the presidency. New York, Pennsylvania, and Illinois were committed to his support by the influence of their powerful Republican leaders; but not unanimously. The movement is supposed to have been undertaken without consultation with Grant; but he did nothing to discourage it, and to this extent he consented to it. The attempt failed. Prudent people had no mind to have their hero's good name again made opprobrious by fresh scandals, which they could not but dread.

CHAPTER XIX

REVERSES OF FORTUNE — ILL HEALTH — HIS LAST VICTORY — THE END

General Grant now made his home in the city of New York. He was not wealthy, and he desired to be. The only persons he seemed to envy, and particularly to court, were those who had great possessions. He coveted a fortune that should place his family beyond any chance of poverty. This weakness was his undoing. He became the private partner of an unscrupulous schemer and robber, and intrusted to him all that he had, and more, to be adventured in speculation. His name was dishonored in Wall Street by association with a scoundrel whom prudent financiers distrusted and shunned. He was warned, but would not heed the warnings. The charitable view is that he was deceived by repayments which he was told were profits. On May 6, 1884, a crisis came and Grant was ruined.

He gave up everything he possessed in the struggle to redeem his honor, even the presents and trophies which had been lavishly bestowed upon him. This savior of his country and recipient of its grateful generosity, who was but lately the guest of the princes of the earth, became dependent upon pitying friends for shelter and bread, until enterprising editors of magazines began competing for contributions from his pen.

And, as if his misfortunes were not yet sufficiently desperate, illness came. A malignant, incurable cancer appeared in his mouth. He stood face to face with the last enemy, the always victorious one, and realized that the rest of life was but a few months of increasing torture. Then the magnificent courage of his soul asserted itself in fortitude unequaled at Donelson, or Vicksburg, or Chattanooga, or the Wilderness. No eye saw him quail ; no ear heard him complain.

It was suggested that if he would write a book, an autobiographical memoir, the profit of it, doubtless, would place his family above want. Nothing can be imagined more unacceptable to General Grant's native disposition than the narration for the public of his own life story. But in his circumstances, the question was not one of sentiment, but only of duty to those who were dependent upon him. The task was undertaken resolutely, and, in spite of physical weakness and suffering, was carried on with as high and faithful energy as he had shown in any campaign of the war. On March 3, 1885, he was restored to the army with the rank of general on the retired list with full pay. He was glad; but in his feebleness joy was as hard to bear as grief. He began failing more rapidly.

In June he was taken to the sweet tonic air of a cottage on Mount McGregor, near Saratoga. Here, in pleasant weather, he could sit in the open air and enjoy the agreeable prospect. But whether indoors or out, he toiled at the book in every possible moment, writing with a pencil on tablets while he had strength, then dictating in almost inaudible whispers, little by little, to an amanuensis.

So, toilsomely, through intense suffering, sustained by indomitable will, this legacy to his family and the world was completed to the end of the war. His last battle was won. Four days after the victory, he died, July 23, 1885. The book had a success beyond all sanguine expectations, and accomplished the purpose of its author. To his countrymen it was a revelation of the heart of the man, Ulysses Grant, in its nobility, its simplicity, and its charity, that has endeared him beyond any knowledge afforded by the outward manifestations of his life.

His conversations in his last days, as reported by visitors to Mount McGregor (among these was General Buckner, who surrendered Fort Donelson), show a soul serene and cheerful, devoted to his country, to humanity, and to peace. No experiences of malevolence and injury had shaken his trust in the goodness of the great majority of mankind.

When the great soldier died he owned no uniform in which he could be suitably attired for the grave, no sword to be laid on his coffin. His body lies in the magnificent tomb, erected by the voluntary contributions of admiring citizens, the commanding attraction of a beautiful park overlooking the broad Hudson as it sweeps past the nation's chief city. Already this resting place has become a veritable shrine of patriotism. Military and naval pageants make it their proper goal, as when, after Santiago, the returning battleships moved in stately procession up the Hudson to the tomb of our national military hero, there to thunder forth the triumphant salute, like a summons to his spirit to bestow an approval.

The Riverside Press
Electrotyped and printed by H. O. Houghton & Co.
Cambridge, Mass., U. S. A.

Made in United States
Cleveland, OH
05 November 2026

47645844R00105